Praise for *If Science i.* T0282853

'Reading this book is like cosying up to a fireside chat where one of the greatest minds in science distills the complex interface between science and the welfare of society. Even better, no fire required!'

Marcia K. McNutt, President of the National Academy of Sciences

'This is a powerful humane argument for science. Martin Rees draws on his long and wide range of experience to show how science works – and how it can be done better.'

David Willetts, President of the Resolution
Foundation and Chair of the UK Space Agency

'Sometimes it can feel like we're stumbling from one global challenge to another. At the same time, we rightly worry about the pace of technological advances. But we cannot afford to turn our backs on science for, without a scientific understanding of our world, we are doomed. Rees puts the case for placing our trust in science compellingly and with a rare honesty.'

Jim Al-Khalili, University of Surrey and BBC Broadcaster

'This timely and absorbing book issues a clarion call to scientists, policy makers, and citizens everywhere to join forces so that the extraordinary advances in science will be directed towards solving pressing global challenges. Whether we live in the best of times or the worst of times in the future is up to all of us.'

Shirley M. Tilghman, President Emeritus of Princeton University

'The future of humankind depends on science and on fully integrating that science into human culture and society. In this erudite yet accessible book, Martin Rees makes a compelling case for supporting science and making it an integral part of our democracy and political decision making.'

Paul Nurse, Nobel Laureate, Director and
Chief Executive of the Francis Crick Institute

'Delight along with me as Martin Rees describes his world of science and what it will take to ensure that we not only survive but prosper in the Anthropocene Era.'

Charles F. Kennel, Scripps Institution of Oceanography
and former Chair of the NASA Advisory Council

'This lucid and compelling book by one of the world's foremost and far-seeing scientists shows why we ignore science at our peril. The book should be required reading for scientists and is an accessible "must read" for everyone interested in the critical and existential challenges facing humanity.'

Ian Goldin, Director of the Oxford Martin School,
University of Oxford

'Martin Rees is unique in combining achievement at the very highest echelons of science, an almost cosmic perspective on humanity's risks and prospects, and the ability to communicate complex ideas in vivid ways. Here he shows how much we need science, in a tour rich in personalities and history, fusing comments on the frontiers of science with appreciation of their political and ethical dilemmas, giving the reader the pleasure of learning in the company of a sparkling intellect.'

Geoff Mulgan, University College London, and former Chief Executive
of the National Endowment for Science, Technology and the Arts

'Are science and its organization fit for purpose as society faces 21st-century challenges, from climate change to dominance by AI? As clearly demonstrated in this masterpiece, no one is better able to answer this crucial question than Martin Rees.'

Tim Palmer, Royal Society Research Professor, University of Oxford

If Science is to Save Us

If Science is to Save Us

MARTIN REES

polity

Copyright © Martin Rees 2022

The right of Martin Rees to be identified as Author of this Work has been asserted in accordance with the UK Copyright, Designs and Patents Act 1988.

First published in 2022 by Polity Press

Polity Press
65 Bridge Street
Cambridge CB2 1UR, UK

Polity Press
111 River Street
Hoboken, NJ 07030, USA

All rights reserved. Except for the quotation of short passages for the purpose of criticism and review, no part of this publication may be reproduced, stored in a retrieval system or transmitted, in any form or by any means, electronic, mechanical, photocopying, recording or otherwise, without the prior permission of the publisher.

ISBN-13: 978-1-5095-5420-1

Library of Congress Control Number: 2022934662

A catalogue record for this book is available from the British Library.

Typeset in 11 on 15pt Warnock Pro
by Cheshire Typesetting Ltd, Cuddington, Cheshire
Printed and bound in Great Britain by TJ Books Ltd, Padstow, Cornwall

The publisher has used its best endeavours to ensure that the URLs for external websites referred to in this book are correct and active at the time of going to press. However, the publisher has no responsibility for the websites and can make no guarantee that a site will remain live or that the content is or will remain appropriate.

Every effort has been made to trace all copyright holders, but if any have been overlooked the publisher will be pleased to include any necessary credits in any subsequent reprint or edition.

For further information on Polity, visit our website:
politybooks.com

Contents

Preface and Acknowledgements

My prime thanks are due to my editor, Jonathan Skerrett. He persuaded me (and his colleagues) that an accessible work on the themes of this book would be appropriate for Polity Press, and he invested a great deal of time and effort into shaping the book and offering helpful suggestions on both style and content. I'm most grateful also to Ian Tuttle for his careful copy-editing and for suggesting numerous improvements, as well as to Neil de Cort and Emma Longstaff, who helped to see the book smoothly through to publication.

It will be plain from the book's contents – spanning wide-ranging themes where I can't claim special expertise – that I owe a lot to what I've learnt, and advice I've received, from a huge number of friends and colleagues with whom I've collaborated or interacted. I cannot mention them all, but I wish to offer special thanks to Partha Dasgupta, Mario Livio and Steven Pinker, whose collaboration has influenced my coverage of some topics.

I also thank the BBC for permission to include in Chapter 2 some updated material adapted from my 2010 Reith Lectures on 'Scientific Horizons'.

Introduction

In our responses to Covid-19 we were told to 'follow the science' – and there's never been a time when 'experts' have had such public prominence. At the time of writing, this pandemic is still, after more than two years, an overwhelming challenge. But it's not the only one: politicians need also to confront a whole array of policy issues – on energy, health, environment, and so forth. Indeed, the choices our governments make in the coming decades could determine the Earth's future. Politicians must take cognizance of expert advice, but in reaching decisions this advice must be tensioned against other factors: the feasibility and public acceptability of particular measures, and their economic and human costs.

Such choices – decisions on how science is applied – should be preceded by an informed public debate. But for the debate to rise above the level of tabloid slogans, we all need a 'feel' for the key ideas underlying modern technology, and an understanding of the natural world (including humans). Science isn't just for scientists. Equally important, we need to be mindful of how incomplete and provisional our knowledge actually is. Moreover, these ideas are not only the basis of everyday technology, but they're of sufficient intrinsic interest that they should be part of our common culture. The great concepts of science – or at least the flavour of them – can be conveyed using non-technical words and simple images. Or so I believe.

In this book I won't elaborate on the actual findings of science. Nor will I extol it as the greatest collective achievement of humanity – though it surely is. My focus will instead be on how the sciences impinge on our lives – and on the hopes and fears for the future. I shall offer thoughts on what distinguishes science from other intellectual activities, how the entire scientific enterprise is organized – nationally and globally – and on how to ensure that scientists and their innovations mesh into society, so that applications are channelled in accordance with citizens' preferences and ethical judgements.

The stakes have never been higher. The Earth has existed for 45 million centuries, but this is the first century in which one dominant species can determine, for good or ill, the future of the entire biosphere. Over most of history, the benefits we garner from the natural world have seemed an inexhaustible resource; and the worst terrors humans confronted – floods, earthquakes and diseases – came from nature, too. But we're now deep in what some have called the 'Anthropocene' era. The human population, now approaching 8 billion, makes collective demands on energy and resources that aren't sustainable without new technology and threaten irreversible changes to the climate. The threat of nuclear war still looms over us. And other novel technologies – especially bio and cyber – are socially trans-formative but open up the possibility of severe threats if misapplied. The worst threats are no longer 'natural' ones: they are caused (or at least aggravated) by humanity itself. There remains a huge – and widening – gap between the way the world is and the way it could be. Inequalities within countries, and between countries, are vast.

Despite the concerns, there are powerful grounds for optimism. For most people in most nations, there's never been a better time to be alive, thanks to advances in health, agriculture and communication – dependent on earlier scientific discoveries – which have boosted the develop-ing as well as the developed world. And this optimism need not be eroded by the pandemic. Indeed, in dealing with this globe-spanning plague, science is our salvation: the response has shown the scientific community at its best – a colossal worldwide effort to develop and deploy vaccines, combined with honest efforts to keep the public informed.

Creativity in science and the arts is nourished by a wider range of influences than in the past and is accessible to hugely more people worldwide. We're embedded in a cyber-space that can link anyone, anywhere, to all the world's information and culture, and to most other people on the planet. Everyday life has been transformed in less than two decades by mobile phones, social media, and the internet – we would have been far less able to cope with the recent shutdowns without these facilities. Computers double their power every two years. Gene sequencing is a million times cheaper than 20 years ago: spin-offs from developments in genetics could soon be as pervasive as those we've already seen from the microchip.

These rapid advances, and others across the whole of science, raise profound questions. Who should access the 'read-out' of our personal genetic code? How could lengthening lifespans affect society? Should we build nuclear power stations, or wind farms, if we want to keep the lights on? Should we use more insecticides, or plant GM crops? Should the law allow 'designer babies'? How much should artificial intelligence (AI) be permitted to invade our privacy? Are we prepared to accept a machine's decisions on issues that matter to us?

All these questions require engagement of 'experts' with politicians and the wider public. The public and governmental challenges posed by the Covid-19 crisis were unprecedented (at least in peacetime) in their urgency, impact and global scope. Some threats – global pandemics and massive cyber attacks, for instance – are immediately destructive and could happen at any time. The worst of them could have consequences that cascade and spread devastatingly. And their probability and potential severity is increasing. Covid-19 must be a wake-up call, reminding us – and our governments – of our vulnerabilities.

Looming over the world this century is the threat of climate change. This is potentially a 'global fever', in some ways resembling a slow-motion version of Covid-19. For instance, both crises aggravate the level of inequality within and between nations. Those in the megacities of the developing world can't isolate from rogue viruses; their medical care is minimal, and they're less likely to have access to vaccines. And likewise, it's those countries, and the poorest people in them, that will suffer most from global warming and the effects on food production and water supplies. Climate change and environmental degradation may well, later this century, have global consequences that are even graver than pandemics – and longer term (indeed irreversible). But a potential slow-motion catastrophe doesn't engage public and politicians – our predicament resembles that of the proverbial boiling frog, contented in a warming tank until it's too late to save itself. We're well aware of them, but fail to prioritize countermeasures because their worst impact stretches beyond the time horizon of political and investment decisions. Politicians recognize a duty to prepare for floods, terrorist acts and for other risks that are likely to materialize in the short term – and are localized within their own domain. But they have minimal incentive to address longer term threats that aren't likely to occur while they're still in office – and which are global rather than local.

The case for effective action to address long-term threats is compelling. But unless there's a clamour from voters, governments won't properly prioritize measures crucial for future generations. So scientists must enhance their leverage, by involvement with non-governmental organizations (NGOs), via blogging and journalism, and enlisting charismatic individuals and the media to amplify their voice and change the public mindset. It's encouraging to witness more activists – especially among the young, who can hope to live into the twenty-second century. Their campaigning is welcome. Their commitment gives grounds for hope.

Without earlier scientific insights, we'd be denied all the everyday benefits whereby our lives differ from those of our forebears – electricity, vaccines, transport and information technology (IT). We should be evangelists for new technology, not luddites – it's essential if the world's expanding and more demanding population is to have enough food and

enough energy in a sustainable form. But many are anxious
that it's advancing so fast that society may not properly
cope with it – and that we'll have a bumpy ride through this
century.

And, of course, most of the challenges are global. Coping
with Covid-19 is plainly a global challenge. And the threats
of potential shortages of food, water and natural resources –
and transitioning to low-carbon energy – can't be solved by
each nation separately. Nor can the regulation of potentially
threatening innovations – especially those spearheaded by
globe-spanning conglomerates. Indeed, a key issue is to
what extent, in a 'new world order', nations will need to
yield more sovereignty to new organizations along the lines
of the International Atomic Energy Agency, World Health
Organization, etc.

Scientists have an obligation to promote beneficial applications of their work in meeting these global challenges.
Their input is crucial in addressing the downsides: helping governments to decide wisely which scary scenarios
– eco-threats, or risks from misapplied technology – can
be dismissed as science fiction and how best to avoid the
serious ones. And we need the insights of social scientists
to help us envisage how human society can flourish in a
networked and AI-dominated world.

My own research field is astronomy and cosmology.
Before concluding this introduction it's perhaps appropriate
to ask: are there special perspectives that astronomers can
offer to this book's theme? I think there are. Astronomers
are disclosing insights that New Agers would welcome and
be attuned to. Not only do we share a common origin, and
a common 'genetic code', with the entire web of life on
Earth, but we are linked to the cosmos. All living things are

energized by the heat and light from the nearest star, our Sun; and the atoms that we – and indeed our entire Solar System – are made of were forged from pristine hydrogen, billions of years ago, in faraway stars.

But, more significantly, astronomers can offer an awareness not only of the immensity of space but of the immense time spans that lie ahead. The stupendous time spans of the evolutionary past are now part of common culture (apart from in creationist circles). We and our biosphere are the outcome of about four billion years of evolution. But most people still somehow think we humans are necessarily the culmination of the evolutionary tree. That hardly seems credible to an astronomer, aware of huge time horizons extending into the future as well as into the past. Our Sun formed four and a half billion years ago, but it's got six billion more before its nuclear fuel runs out. And the expanding universe will continue – perhaps for ever – becoming (according to the best current long-range forecast) ever colder, ever emptier. So, even if life were now unique to Earth, there would be scope for post-human evolution – whether organic or electronic – on the Earth or far beyond. It won't be humans who witness the Sun's demise: it will be beings more different from us than we are from a bug. We can't conceive what powers they might have.

This book has of course a narrower theme than the cosmos. The focus is on our Earth, and mainly on the present century: an instant in cosmic perspective, but sadly longer than the planning horizon of business and politics. My focus is on scientists – their communities and their interaction with society, the economy and politics rather than on what their work has revealed about nature. Indeed, I'll be using the term 'science' – as is common practice in public discourse – to embrace technology and engineering as well. 'Problem solving' motivates us all – whether one is an engineer facing a novel design challenge or an astronomer probing the remote cosmos. And having myself had a career focused on academic science, I want to emphasize that, despite their symbiosis with 'pure' science, it's the 'applied' activities that engage far more brainpower and resources. The message of an old cartoon that resonates, quite rightly, with my engineering friends shows two beavers looking up at a giant dam. One beaver is saying to the other 'I didn't actually build it, but it's based on my idea.'

* * *

Chapter 1 highlights three areas where science is transformative – and indeed where the whole future of our species depends on its deployment for societal benefit. These are climate and environment, biomedicine, and computers and machine learning. I argue that science and technology – optimally applied – will be crucial to our collective flourishing. But we need to be mindful of the downside; some technologies are advancing so fast that we may not properly cope with them – and their misuse, by error or by design, can lead to catastrophe. There will always be a trade-off between risks and benefits, and it's therefore important that public concerns are respected, and that these aren't distorted by unbalanced perceptions.

Chapter 2 describes what scientists are like – emphasizing that rather few actually resemble traditional stereotypes, in personality or work patterns – and how their ideas are communicated, to become part of our culture as well as the underpinning of our modern (and future) world. I address the structure and sociology of the scientific enterprise;

science's scope and limits, and its relation to culture and politics; and how to improve the public's capacity to make informed choices of how it is applied. Scientists must acknowledge that the applications of their work resonate far beyond their expertise; citizens and politicians must be reassured that new discoveries aren't applied unethically or dangerously.

In chapter 3, I describe the institutions within which scientists work – some of which have serious weaknesses. Nations differ in the extent to which their scientists can engage with their governments as advisors, or directly with the public via campaigning and the media. The role of international organizations and academies needs strengthening, especially as the challenges we face increasingly require a coordinated international response. Science is a truly global culture, and we need deeper international contacts among professionals, and in universities and colleges.

Being a scientist is a career choice – it's crucial that enough talented people should opt for this choice. Those that do require sufficient incentives and appropriate education and opportunities. So chapter 4 addresses educational issues, not only from the perspective of potential professionals, but in the wider context of ensuring that all of us understand enough to feel at home in our high-tech world and can participate in debates on how science is applied. Formal education – throughout the school years, and in higher education as well – is one of the most sclerotic aspects of UK society; the US offers greater flexibility, but at school level the whole Anglo-Saxon world can learn from Scandinavia and the Far East. The world is changing so fast that learning must be a lifelong process: it needs to be inclusive and flexible, not restricted to a privileged minority; it should take optimum advantage of the internet. Some words of H. G. Wells a century ago resonate even more strongly today: we're in 'a race between education and catastrophe'.

Chapter 1

Global Mega-challenges

The 'plague years' of Covid-19 have imprinted two contrasting messages. First, our entire world is interconnected: a catastrophe in any region can cascade globally; no nation is truly safe until all are. Second, international science can be our salvation – as in the development of vaccines. Let's hope that, when this crisis has passed, nations can focus on ensuring that we're better prepared for the next pandemic. Moreover, it has been a 'wake-up call' that should deepen concern about other future threats that could be even more catastrophic; it should stimulate effective actions to confront all the longer-term challenges the world faces.

I'd highlight three interlinked mega-challenges:

1. Providing food and energy for a rising and more demanding population, while avoiding depletion of the biosphere and dangerous climate change.
2. Coping with the ethical and security challenges posed

by ever-advancing biotechnology while harnessing its
benefits for health and agriculture.
3. Enabling artificial intelligence, the cybernet and social
media to transform our economy and our society, despite
vulnerability to malfunctions (natural or malicious) that
could cascade globally.

The alarm having sounded, these are items which have long
been on humanity's collective agenda but which we now
need to consider anew.

1.1 Threats to the biosphere: population growth and biodiversity loss

The backdrop to current geopolitical challenges is a world
where humanity's collective footprint is getting heavier.
There are about 7.8 billion of us on this planet – twice
as many as in the 1960s. Nonetheless, despite doom-laden
forecasts by Paul Erlich (1968)[1] and the Club of Rome
(1972),[2] food production has, thanks largely to advances
in plant science (the 'green revolution'), kept pace with
rising population. Famines still occur, and many people,
especially children, remain undernourished; but the most
distressing episodes, such as those in Afghanistan, Yemen
and Ethiopia, are mainly due to conflict or maldistribution,
not overall scarcity.

Population growth has now slowed. Indeed, the number
of births per year, worldwide, is now declining: in most
countries it has fallen below the 'replacement level' of 2.1
births per woman; for instance, it is 1.5 in Japan, 1.56 in
Canada, and 1.64 in China, leading to concerns (especially

in Japan) about an over-dominance of the elderly. But world population is nonetheless forecast to rise to around 9 billion by 2050.[3] That's partly because most people in the developing world today are young, owing to persistent high fertility in recent decades and welcome falls in infant mortality. These young people are yet to have children, and they will live longer. Moreover, the transition to low fertility hasn't happened everywhere – particularly in rural parts of sub-Saharan Africa.

Most of the 7.8 billion people on the Earth today are still impoverished by the standards of the 'Global North' – though, according to the World Bank, the proportion below the official 'extreme poverty' threshold, which currently stands at $1.90 per day, has dropped from around 60 per cent in 1950 to 10 per cent today.[4] World food production needs to double again by 2050, not only to cope with the rise in population but to ensure that all those in the Global South (where the main population growth in the coming decades will be) become as well nourished as most people in Europe and North America.

It's true that food production has doubled in the last 50 years; but a further doubling is more problematic. There will be constraints on energy, on the quantity of fertile land, and on the supply of water. This will require further improved agriculture – low-till, water-conserving and genetically modified (GM) crops – together with greater efforts to reduce waste (via refrigeration, for instance) and improve irrigation. We need modes of farming that can produce crops efficiently in a changing climate, and avoid encroaching on natural forests. The buzz-phrase is 'sustainable intensification'.[5] There will be consequent pressure to enhance the yield from the oceans, without allowing over-fishing to drive species to extinction. There will certainly need to be changes in the typical 'Western' diet: for instance, we can't all consume as much beef as present-day Americans.

Some dietary innovations are feasible without deployment of 'frontier' science: for instance, converting insects and maggots into palatable food, and making artificial meat from vegetable protein. 'Beef' burgers (made from wheat, coconut and potato, moisturized with beetroot juice) are now being marketed in the US by companies called Beyond Meat and Impossible Foods. It will be a while, though, before these 'pseudo-burgers' will satisfy carnivorous gourmands.

These novel foods are best characterized as clever or exotic cookery rather than entailing advances in a laboratory. But biochemists are now making breakthroughs that portend more fundamental innovations; they 'grow' meat by taking a few cells from an animal and then stimulating their replication with appropriate nutrients. In 2020, Singapore's food regulatory agency approved the sale of 'cultured' meat developed by a US start-up called 'Eat Just'. Meat substitutes acceptable to fastidious carnivores are clearly an ecological benefit. The concern is whether they can be produced sufficiently cheaply for a mass market. Let's hope so, because for many of us they would be a welcome ethical advance too; future generations may look back in horror and disgust at the 'factory farming' techniques that prevail today.

But it's an open question how readily these seemingly 'un-natural' foods will be accepted, even if they become affordable – it would be dismaying if they were only welcomed as petfood. The history of GM crops is a worrying augury. Despite their clear benefits – and despite the fact that 300 million people in North America have eaten GM crops for two decades without manifest harm – they have been banned in the European Union, where an extreme 'precautionary principle' is adopted. GM crops have even been rejected when sent to Africa as 'food aid' to alleviate famine.

There's a well-known estimate from the World Wide Fund for Nature (WWF) that the world is already despoiling the planet, by consuming natural resources at around 1.7 times the sustainable level.[6] Such considerations (and others) suggest that a decline in world population – provided that it were gradual – would be optimal. But it's important to recognize that it's not feasible to define a definite 'carrying capacity' for the world. The present global population would be massively unsustainable if everyone lived like present-day Americans. (Diets rich in animal products have much higher footprints than those based on plant products. If pastureland and land used for livestock feed are combined, animal agriculture already uses nearly 80 per cent of global agricultural land.) On the other hand, one can imagine – without recommending! – a sustainable science fiction dystopia where 20 billion people could live in capsules, travelling little and contentedly experiencing 'virtual reality'. There are choices to be made every step of the way in seeking to solve such problems.

If humanity's collective impact on nature pushes too hard against what the Swedish environmentalist Johan Rockström calls 'planetary boundaries',[7] the resultant 'ecological shock' could impoverish our biosphere irreversibly. In densely populated countries, there's widespread anxiety about the consequences for wildlife of urbanization, pesticides, and so forth – though there is understandably an undue focus on birds and on charismatic or cuddly mammals rather than on worms, insects and microfauna that are more crucial to a balanced ecosystem. The negative impact of biodiversity loss is still too low on the global agenda. This is partly because those in poor countries, though at the sharp end of its impacts, excusably have a shorter time horizon; and partly because some ecosystems are spatially extensive and straddle many national boundaries. The oceans, as a prime example, remain largely a 'common resource' that will need firmer regulation.

Biodiversity is threatened when land is built on, cultivated or overgrazed – also when large areas are subdivided – and its loss would be aggravated if the extra land for food production or biofuels encroached on natural forests. Changes in climate and alterations to land use can, in combination, induce 'tipping points' that amplify each other and cause runaway and potentially irreversible change.

Diverse ecosystems are more sustainable, more resilient. When conditions change, some minority species with different traits may get an advantage. They are, as it were, 'waiting in the wings', to take over if required to do so. Sparser ecosystems can't respond so well to changing conditions – rather as a sports team with substitutes on the bench is best placed if there's a range to choose from with different skills. To quote the iconic environmentalist David Attenborough:

> Today, we ourselves, together with the livestock we rear for food, constitute 96% of the mass of all mammals on the planet. Only 4% is everything else – from elephants to badgers, from moose to monkeys. And 70% of all birds alive at this moment are poultry – mostly chickens for us to eat. We are destroying biodiversity, the very characteristic

that until recently enabled the natural world to flourish so abundantly. If we continue this damage, whole ecosystems will collapse. That is now a real risk. Putting things right will take collaborative action by every nation on earth. It will require international agreements to change our ways. Each ecosystem has its own vulnerabilities and requires its own solutions. There has to be a universally shared understanding of how these systems work, and how those that have been damaged can be brought back to health.[8]

But the main impediment to global-scale implementation of Attenborough's injunction is that even though the natural world is crucial to us all, 'natural capital' doesn't feature in national budgets. If, for example, a forest is cut down, it should be recorded as a negative contribution to a nation's stock of natural capital. This has long been urged by the development economist Partha Dasgupta – a colleague of mine at Cambridge – but currently, in most countries, it does not happen. His 500-page report[9] published in 2021, the UK's input to the UN Biodiversity Conference in Kunming, China in 2022, is an impressive document. It deserves as much attention as Nicholas Stern's celebrated report on the economics of climate change, published back in 2006.[10] As Dasgupta explains:

Being embedded within [the biosphere], we are entirely dependent on it, not just for survival but for our well-being too. Nature's goods and services are the foundations of our economies. They include the provisioning services that supply the goods we harvest and extract (food, water, fibres, timber, medicines) and cultural services, such as the gardens, parks and coastlines we visit for pleasure, even emotional sustenance and recuperation. But Nature's processes also maintain a genetic library, preserve and regenerate soil, control floods, filter pollutants, assimilate waste, pollinate crops, maintain the hydrological cycle, regulate climate, and fulfil many other functions besides. Without those regulating and maintenance services, life as we know it would not be possible.

The most devastating consequence of biodiversity loss is extinction—destroying the book of life before we've read it. Biodiversity is crucial to human wellbeing – essential for our health and prosperity. We are all deprived if over-fishing leads some species to extinction; the rainforests host plants whose gene pool may have medicinal use. But life's diversity has a spiritual value, too, eloquently expressed by the great ecologist E. O. Wilson:

Natural ecosystems – forests, coral reefs, marine blue waters – maintain the world as we would wish it to be maintained. Our body and our mind evolved to live in this particular planetary environment and no other.[11]

These sentiments resonate with conservationists: many would indeed take the ethical stance that preserving the richness of our biosphere has value in its own right, over and above what it means to us humans. To quote E. O. Wilson again: 'mass extinction is the sin that future genera-tions will least forgive us for'.

So, what about the more distant future? Population projections beyond 2050 are uncertain – it's not even clear whether there'll be a continuing global rise or a turn-around. If, for whatever reason, families in large parts of

Africa remain large, then, according to UN projections, that continent's population could double again by 2100, to 4 billion, thereby raising the global population to 11 billion. The population of Nigeria alone could then rival that of North America and Western Europe combined.

But could African nations then escape the poverty trap – let alone close the economic gap with the 'Global North'? It's the geopolitical stresses – the inequalities between countries as well as within countries – that are the most worrying aspects of this scenario. As compared to the fatalism of earlier generations, those in poor countries now know what they're missing: they may not have toilets, but their access to IT gives them a window on the world that highlights the injustice of their fate. And migration is easier. It's a portent for disaffection and instability. Wealthy nations, especially those in Europe, should urgently promote growing prosperity in poor nations of the 'Global South', and not just for altruistic reasons. There needs to be a 'mega-Marshall' or 'Lend-Lease' plan for sub-Saharan Africa – a region seemingly gaining more from the Chinese 'Belt and Road' initiative than from the West.

A global challenge will be to feed 9 billion people in 2050, without unduly encroaching on natural habitats or threatening biodiversity. 'Food science' may seem an unglamorous topic. But developing intensive agriculture, eliminating plant diseases – and even creating entirely new foods, provided that they are culturally and gastronomically acceptable – is a prerequisite for sustaining health and prosperity. A country like the UK hosts only 1 per cent of the world's population, and encompasses an even smaller fraction of its land mass. Nonetheless, by sharing their expertise in plant science and genetics, and by properly deploying this expertise, the technically advanced countries of the North can have disproportionate leverage in promoting global sustainable development. This was defined in Gro Harlem Brundtland's classic 1987 report[12] as

> Meeting the needs of the present – especially the poor – without compromising the ability of future generations to meet their own needs.

And they can achieve this without degrading the wonders and beauty of the natural world. This is an inspiring challenge for the younger generation – and an investment in the future.

Similar arguments apply to the even greater challenge of generating 'clean energy' for the world – a topic to which I now turn.

1.2 The climate and energy crisis

The world is getting more crowded. And there's another firm prediction: it will get warmer. In contrast to food and population issues, climate change is certainly not under-discussed, though it has not had the response it demands. And it displays 'writ large' the tensions that can arise between scientists, the public, and policymakers, which are the themes of this book.

Climate science is complex, involving a network of intermeshing effects. But there is one fundamental piece of evidence: the amount of carbon dioxide in the atmosphere is higher than it's been for 2 million years, and it is inexorably rising, mainly because of the burning of fossil fuels. This measurement isn't controversial; the best-known evidence comes from the so-called 'Keeling curve', based on measurements made in Hawaii by Charles and Ralph Keeling, father and son, over a period of more than 50 years.[13] Straightforward chemistry, already understood in the nineteenth century, tells us that carbon dioxide is a 'greenhouse gas': it acts like a blanket, preventing some of the heat radiated by the Earth (as infrared radiation) from escaping freely into space. So the carbon dioxide build-up in the atmosphere will cause an inexorable long-term warming, superimposed on other effects – like the El Niño phenomenon in the South Pacific – that generate fluctuations on timescales of up to a decade. Moreover, the science is complicated by 'feedbacks' which can enhance the direct warming effect of CO_2. In particular, a warmer atmosphere can retain more water vapour – which is itself a 'greenhouse gas' but which also affects the cloud cover; this will in turn affect the pattern of winds and rainfall. Moreover, the single figure that is normally used to relate global warming to CO_2 concentrations is just an overall average: the temperature rise will be uneven – different over land and sea, and at different latitudes.

In August 2021 the International Panel on Climate Change (IPCC) published the sixth report of its 'Working Group 1', which presented a spread of climatic projections for different assumptions about future rates of fossil fuel use. The IPCC is a sociologically impressive collective enterprise whereby scientists from around the world collaborate to distil a prognosis for the world's climate and its implications for policymakers (a big contrast in scientific style from the Keelings' individualist efforts). It was established by the UN in 1988 and involves most of the world's leading climate experts. Divided into three working groups and a taskforce, each group has a chair from a developed country and a chair from a developing country. A new 'state of the world' assessment report is issued every seven years or so. Scientists review and synthesize developments in monitoring and understanding climate change impacts, adaptation, vulnerability and mitigation. 'Working Group 1'[14] is focused on climate science, and relied on more than 14,000 peer-reviewed studies from around the world. The two drafts of the report received over 70,000 comments from experts and was signed off by scientists from 66 countries. It was then reviewed – as it was commissioned by the UN – by representatives of 195 member governments. There were three major findings: 1. Humans are 'unequivocally' responsible for global warming (obvious to many, but the IPCC's strongest statement yet); 2. Some climate-induced changes, such as continued sea level rise, are irreversible at least for centuries; 3. It is very late, but thankfully not too late, to avoid the worst impacts of climate breakdown.

The Earth is warmer than it's been since before the last Ice Age (125,000 years ago). The Working Group said it was 'more likely than not' that the world would breach 1.5°C above pre-industrial levels by 2040 – a temperature rise that it claims would be likely to bring 'multiple interrelated climate risks'. Some Pacific Island nations may no longer exist by 2100. The 42-page summary for policymakers included, according to *The Washington Post*,[15] more than 100 uses of the words 'high confidence' and nearly a dozen instances of 'virtually certain'. The IPCC is seen by some scientists as being diplomatically measured and cautious in its pronouncements: climate change has often outpaced its projections, so terms like these aren't thrown around flippantly: there is a far firmer and indeed more alarming tone than in the fifth report released in 2013.

Compared with earlier IPCC assessments, the latest one is based on a longer stretch of data; this firms up evidence for long-term trends, yielding a compelling case that mean temperatures have risen about 1.1°C higher than in the nineteenth century. Moreover, climate change is already manifested by more frequent extreme heatwaves and rainstorms. (There was a cascade of extreme events in 2021; record temperature in Siberia and the West Coast of North America; and catastrophic flooding in Germany and neighbouring countries.) The latest report of Working Group 1 narrows down the range of uncertainty in the projections. The modelling is improving in two main ways. First, more powerful computers allow a finer 'mesh', which can simulate more reliably the complex patterns of wind and temperature in the atmosphere. Second, the way clouds form is becoming better understood. (To mention one subtlety, water vapour in the upper atmosphere condenses into droplets. These cool below 0°C before freezing into crystals, and the extent of this 'supercooling' determines how readily clouds form.)

The warming is much faster in the Arctic – it was previously claimed to be twice as fast; but recent claims suggest that in the last two decades it has actually been four times faster. Ice reflects (rather than absorbs) sunlight to a greater extent than open ocean, so the reduction of Arctic ice exerts 'positive feedback' that accelerates the overall warming; by destabilizing the 'polar vortex' and jet stream, it leads to more extreme climatic variations at intermediate latitudes. More worryingly, the modelling of unmitigated 'business as usual' scenarios suggests that we can't rule out, later in the century, really catastrophic warming, an acceleration of long-term trends like the melting of ice in Greenland and the Antarctic, with consequent continuing sea-level rises.[16]

So much is at stake that it's crucial to develop a still better understanding so that predictions can be firmed up. Nonetheless, even the existing science convinces most experts that the risk of seriously disruptive climate change is growing year by year – and is threatening enough to justify its priority on the political agenda. The reports of the IPPC's other two Working Groups – dealing with 'impact, adaptation and vulnerability' and with 'mitigation' – appeared in early 2022 and highlight the urgency of the need to shift away from a carbon-fuelled economy.

The confidence and urgency implicit in these reports may surprise anyone who dips at random into all that's been written on the subject. Trawls of the internet reveal diverse and contradictory claims. How do you make up your mind? The following analogy suggests an answer. Suppose you seek medical guidance. Googling any ailment reveals a bewildering range of purported remedies. But, if your own health were at stake, you wouldn't attach equal weight to everything in the blogosphere: you'd entrust yourself to someone with manifest medical credentials and a successful record of diagnosis. Likewise, we get a clearer steer on climate – though not, of course, a complete consensus – by attaching more weight to those with serious credentials in the subject, whose views are incorporated in the IPCC reports.

It's still crucial, however, to keep clear water between the science on the one hand, and the policy response on the other. Risk assessment should be separate from risk management. So even those who accept the IPCC's scientific predictions, and believe there's a significant risk of climate catastrophe within a century, differ in how urgently they advocate action today. These divergences stem from differences in economics and ethics – in particular, in how much obligation we should feel towards future generations: whether we discount long-term concerns or feel we should, as it were, pay an insurance premium to reduce the risk to our descendants. Another ethical issue is whether the nations that have been the major 'polluters' in the past have an obligation to those of the Global South who will be far more vulnerable than countries like the UK and US to the consequences of climate change even though they have contributed far less to CO_2 emissions.

But – to insert a bit of good cheer – science can offer a 'win–win' roadmap to a low-carbon future. Nations should accelerate R&D into all forms of low-carbon energy generation; and into other technologies where parallel progress is essential – especially storage (batteries, compressed air, pumped storage, hydrogen, etc.) which is crucial as a complement to unsteady generation via sun or wind. Long-range low-loss grids should also be considered – to bring solar energy from North Africa and Spain to less-sunny northern Europe, and to smooth over the peak consumptions (normally around 7pm) in different time zones by east–west transmission lines – perhaps all the way along the Belt and Road to China. Achieving this would require vision, commitment and public–private investment on the same scale as the building of railways in the nineteenth century. Indeed, complete transformation of the world's energy and transport infrastructure is a massive project that will inevitably take several decades.

Wind and solar energy generation are of course both being widely deployed, though there are 'niches' that can be filled by geothermal, hydro and tidal power. (And, incidentally, tidal energy is a credible option for Britain, as the tidal range along the west coast, especially in the Severn Estuary, is more than 12 metres for spring tides – a level surpassed in the Bay of Fundy in Eastern Canada, but matched in few other locations.)

And what about nuclear power? Despite current ambivalence about widespread nuclear energy, it's surely worthwhile to boost R&D into 'fourth generation' nuclear power – especially small modular reactors – which could be more flexible, and safer, than existing reactors, and indeed cheaper if all follow a standard design. The number of nuclear power stations being built in Western countries has plummeted in the last 20 years. Current designs date back to the 1960s, but have become immensely costly because of extra 'bells and whistles' added to enhance safety. Countries like France that now depend heavily on nuclear energy will need to replace existing power stations; replacements are scheduled for the 2030s (and indeed several in the UK have faults that require earlier decommissioning).

And of course, nuclear fusion, the process that powers the Sun, still beckons as an inexhaustible source of energy. Attempts to harness this power have been pursued since the 1950s. There have been some 'false dawns', but the history is of receding horizons: commercial fusion power still seems at least thirty years away. Most prototypes deploy magnetic forces to confine gas at a temperature of millions of degrees – as hot as the centre of the Sun. Despite its cost and the formidable challenges it poses, the potential pay-off is so great that it is surely worth continuing these developments. By far the most ambitious is the International Thermonuclear Experimental Reactor (ITER), internationally funded and based in France. Similar but smaller projects are being pursued in Korea, the UK, the US and elsewhere, with funding from private investors. Nearly all of these, like ITER, involve 'magnetic confinement' of ultra-hot gas but can be scaled down because they deploy stronger magnetic fields. An alternative concept, whereby tiny deuterium/tritium capsules are zapped and imploded by converging beams from immense lasers, is being pursued at the Livermore Laboratory in the US, but this seems primarily a defence project to provide lab-scale substitutes for H-bomb tests, the promise of controlled fusion power being a political fig leaf. This experiment has recently come close to 'break-even' – in the sense that the thermonuclear power released almost equals the power used to fire up the lasers – but it would seem hard to scale up this process economically.[17]

Real breakthroughs are needed in energy generation, storage and smart grids. And there's a broader and even more compelling motivation for prioritizing such efforts in technically advanced countries, and it's this. Under 'business as usual', the main future rises in annual CO_2 emissions may come from countries in South Asia and Africa, which can't reach acceptable living standards without generating more power than they do today. Not only will their per capita energy needs rise, unlike those in the prosperous North, but they will collectively harbour a billion more people by 2050 – four billion rather than three. 'Bending the trajectory' of CO_2 emissions from these nations is crucial. They must be economically and technically enabled to leap-frog to clean energy rather than building coal-fired power stations (rather as they have transitioned to smartphones without ever building landlines). Technically advanced countries could thereby catalyse a far greater reduction in global emissions than they would achieve just by achieving 'net zero' themselves.

The climate change crisis is potentially a threat to national security. For that reason, combating it deserves the scale of sustained effort that we commit to our national defences – and the focus appropriate to a national emergency. The US is a laggard in making a formal pledge, but it, too, will face a security challenge that will require large-scale long-term mission-driven efforts. In the US, the two successive energy secretaries in the Obama administration – Stephen Chu and Ernie Monoz, both world-class scientists – advocated establishing new national laboratories to spearhead energy innovation, along the lines of the now-defunct Bell Laboratories and those run by the Defense Department. That's what other technically advanced nations need too: institutions with long-term missions devoted to a national goal which are crucial intermediaries between product-driven research in industry, and journal-driven research in universities.

The optimum structure and governance of the needed research and development facilities – hubs of expertise to spearhead innovation and development – deserve serious discussion. The main effort and expense would go into 'development', where the basic principles are understood: this is not a role for universities, though liaison with academics would be crucial. A modest fraction of funding should be reserved for 'blue-skies' exploration of speculative ideas – just as the US assigns a small fraction of its budget to the Advanced Research Projects Agency (ARPA). But how much? And how can we best ensure that there is take-up from industry so that long-term economic benefits accrue?

The public spending on clean energy R&D is only about 2 per cent of all R&D spending. Why shouldn't it be at a level closer to medical or military research? I was therefore glad to join with colleagues in campaigning, in the lead-up to the 2015 United Nations Climate Change Conference in Paris, for major countries (especially those in the G20), to commit 0.02 per cent of GDP to coordinated research into solar energy, energy storage and grids. This seems very modest, but would represent a doubling of the current effort – still not enough, but hopefully politically realistic. This led, thanks largely to chemist David King and economist Nick Stern, to a project called 'Mission Innovation' which achieved some of our goals. A similar scheme would now gain greater support, especially as the mindset of the business and finance sector has genuinely shifted.[18]

A key mantra for technically advanced Western nations should be 'If we don't get smarter, we'll get poorer.' With bold organizational reforms to our education (see chapter 4), and measures to promote an innovation culture, countries like the UK could aspire to contributing far more than their pro-rata share to solving global challenges – and to their economic benefit.

The UK has been especially bold in passing a Climate Change Act that commits us to achieving net zero by 2050. This target is daunting. To achieve it, the nation will need not just to decarbonize existing electricity generation from coal and gas, but also to more than double the current amount of electricity that's generated. This extra is required to cater not only for electric cars and trucks and for home heating (currently mainly gas) but probably for electrolytic production of liquid and gas fuels – hydrogen, methane and kerosene – needed for long-distance aviation (where the requisite batteries would be too heavy), and manufacture of steel and cement. It will require novel technology. It would of course require a huge expansion in wind and solar power generation, along with energy storage on a massive scale, and probably a transcontinental grid to smooth over local weather variations and to carry solar energy northward. But the most likely stumbling block to public acceptance is the requirement to replace the boilers used for heating 22 million homes by heat pumps that are much more expensive.

If all nations pledged to achieve net zero by 2050 – and, more importantly, could be expected to fulfil their pledges – then we could sleep easily in the hope that warming would be limited to 1.5°C. But we are far from that happy state. At the COP 26 conference in 2021, two 'big players' pledged to achieve the target, but more slowly: China by 2060; India by 2070. China now contributes 27 per cent of global emissions, more that the US and EU combined. Others, including the US, have still made no commitment. But we must realistically accept that meeting such a target is an especially difficult challenge for the developing world – and it's a 'hard sell' to the public, even in rich countries, if it imposes extra costs. That's why scientific advances that allow cost-cutting in clean energy generation are crucial.

It will indeed be a challenge to keep the public on side in some countries – even their leaders. We need only recall President Trump's response to climate change. Even after shifting from his early claims that it was 'an expensive hoax' and that 'the concept of global warming was created by and for the Chinese in order to make US manufacturing non-competitive', Trump's actions continued to speak louder than his words. Denying the alarming conclusions of thousands of climate studies worldwide, the US administration started the process of withdrawing from the 2015 Paris climate agreement, attempted to freeze fuel efficiency standards imposed on new vehicles, and weakened regulations aimed at limiting carbon emissions. Fortunately, reversal of some of these decisions was among President Biden's first directives. But the amplification of 'fake news' has ramifications far beyond the scientific controversies that relate to the themes of this book. And I fear it will long outlast Trump.

Even if a nation confronts an immediate threat like Covid-19, we've seen the difficulty in gaining public acceptance of recommended measures: vocal and even violent anti-vaxxers have obstructed vaccination in several countries. If citizens are asked to pay more for travel and heating, mainly for the benefit of distant countries decades in the future, acquiescence is far less likely, and in democratic nations resistance may impede effective action.

So I fear it's highly likely that CO_2 concentrations will still be rising in 2050, despite the pledges and rhetoric. There may then be pressure for a 'Plan B'. There are two options, both involving massive implementation of currently untried technologies.[19] One would entail directly extracting excess CO_2 from the atmosphere and sequestering it. This is benign, but hugely expensive. The alternative would be some form of geoengineering – for instance, injecting aerosols into the upper atmosphere to partially block the sunlight reaching the ground. Some might view this as a quick fix that could cool the Earth enough to delay climate change. It would be within the resources of a single nation or even a large corporation. But it would at best be buying time and storing up worse problems for the future. Moreover, our understanding and modelling of the world's climate would need to be much more reliable and detailed before one could be confident of what such an intervention would actually do at a regional level; not all nations would want to 'turn down the thermostat' equally. Such technology – and other options like, for instance, measures to increase the amount of Arctic ice – would open up new pretexts for international litigation. Only the lawyers would benefit. Some scientists may find these projects alluring. So this is an instance where politicians and the public may need to apply the brakes to a grand project: there is some analogy here with suggestions in the 1950s by Edward Teller that nuclear bombs could be 'peacefully' deployed in massive civil engineering projects like canal-building.

In short, technologies exist (or could be plausibly developed) that could enable a full transition to a net zero world. But political and economic realities are likely to scupper either Plan A or Plan B. Realistically, each country's fate after mid-century will depend on how well it can adapt – difficult for all, especially for those most vulnerable to sea-level rises, and those that are already hottest.

I hope my gloomy forecast on the climate front will prove incorrect, but conclude this section with an instance where a scientist displayed extraordinary foresight. It's an extract from a lecture entitled 'Daedalus, or, Science and the Future'[20] that the polymath biologist J. B. S. Haldane gave in Cambridge in 1923. He sets his vision 400 years in the future. This is not, I think, because he expected slow technological change – indeed other parts of his lecture are wildly futuristic (human embryos raised in flasks, and suchlike) – but because he was unaware of potential climate change, and concerned only about what would happen when readily extractable fossil fuels ran out:

Personally, I think that four hundred years hence the power question in England may be solved somewhat as follows: The country will be covered with rows of metallic windmills working electric motors which in their turn supply current at a very high voltage to great electric mains. At suitable distances, there will be great power stations where during windy weather the surplus power will be used for the electrolytic decomposition of water into oxygen and hydrogen. These gasses will be liquefied, and stored in vast vacuum jacketed reservoirs, probably sunk in the ground. If these reservoirs are sufficiently large, the loss of liquid due to leakage inwards of heat will not be great; thus the proportion evaporating daily from a reservoir 100 yards square by 60 feet deep would not be 1/1000 of that lost from a tank measuring two feet each way. In times of calm, the gasses will be recombined in explosion motors working dynamos which produce electrical energy once more, or more probably in oxidation cells. Liquid hydrogen is weight for weight the most efficient known method of storing energy, as it gives about three times as much heat per pound as petrol. On the other hand, it is very light, and bulk for bulk has only one third of the efficiency of petrol. This will not, however, detract from its use in aeroplanes, where weight is more important than bulk. These huge reservoirs of liquified gasses will enable wind energy to be stored, so that it can be expended for industry, transportation, heating and lighting, as desired. The initial costs will be very considerable, but the running expenses less than those of our present system. Among its more obvious advantages will be the fact that energy will be as cheap in one part of the country as another, so that industry will be greatly decentralized; and that no smoke or ash will be produced.

1.3 Biotechnology: hopes, fears and ethical conundrums

We've learnt during the Covid-19 pandemic how crucial it's been to deploy a global effort to develop and mass-produce vaccines – and thereby enhance our resilience.

More generally, advances in biomedicine have vastly improved health and extended lifespans all over the world, including in its poorest countries. Over the past 25 years an average human has gained seven years of life expectancy. The gain is more than just staying alive: the number of healthy years we can expect to enjoy has risen too. Future benefits to humanity could be greater still. Anyone who has lost a loved one to cancer, heart failure or a degenerative disease can appreciate the unfathomable gift of biomedical advances that would have prevented the premature ending of their active lives. Multiply that gift by billions and you can begin to appreciate the possible bounty of progress in our battle against disease.

At the same time, this progress creates vulnerabilities and ethical dilemmas. Most obviously, the benefits are unevenly shared, both within countries and, even more, between rich nations and the impoverished Global South. Reducing these inequalities is surely an imperative. But, sadly, within many countries the trend is now in the wrong direction: there's too much focus on 'diseases of the rich' rather than on infectious diseases. The gulf between what medical science may enable us to do, and what it is prudent or ethical actually to do, will shift, widen and, in many cases, be difficult to cope with.

Concerns about biomedicine have loomed large during the Covid-19 crisis, but they are nothing new. Throughout the history of medicine, people have recoiled at innovations that seemed to go against nature, including vaccination, transfusions, artificial insemination, organ transplants and in-vitro fertilization. The fact that these are unexceptionable today is a reminder that squeamishness at the new is not a reliable guide to what becomes deemed ethically defensible. More recent techniques that are still controversial include stem cell research and mitochondrial transplants (so-called 'three-parent babies').

The cost of sequencing DNA has plummeted dramatically. The decoding of the first human genome, completed in 2003, was quintessential big science – an international project with a $3 billion budget. But the cost of human genome sequencing has fallen to below $1,000, and soon it will be routine for all of us to get sequenced. At the same time, it has become feasible to synthesize genes and even simple genomes from scratch.

However, most people distinguish between an intervention that would remove something harmful, which they welcome, and one that would enhance what we already have, which they fear. Whether or not this difference is morally significant (or even, in many cases, meaningful), the actual prospect of genetic enhancement of humans is, perhaps fortunately, remote. A few genetic diseases, including Huntington's, are caused by a single gene that could be snipped out by the CRISPR-Cas9 gene-editing technique.[21] But most, such as schizophrenia or a susceptibility to Alzheimer's or cancer, are the product of hundreds or thousands of genes, each tweaking the probability of a person having the disease by a tiny amount. This is even truer of traits and talents such as height, intelligence and personality. Only when the DNA and trait profiles of many millions of people are available will it become possible (using pattern-recognition systems aided by AI) to identify desirable combinations of genes. Not until this can be done will 'designer babies' become conceivable (in both senses).

And perhaps not even then. In the 1990s many commentators fretted about the day wealthy parents would be able to insert a gene for intelligence into their unborn children. The scenario is likely to be perceived differently now. It seems likely that it would require inserting several thousand genes into a child; and even though each insertion might possibly increase mental ability by a minuscule amount, because no gene has a single effect, it might also fractionally increase the probability of (for instance) brain cancer or epilepsy. Moreover, even the most heritable diseases display a lot of unpredictable non-genetic variability, as is clear from the extent to which identical twins can have divergent trajectories of life and health.

But, even if most of us are relaxed about the realistic genetic interventions I've mentioned, there are genuine concerns about where advancing genetics could take us. (Indeed, there is widespread reluctance to properly discuss the issue, because it seems redolent of the now anathematized eugenic movements in the first half of the twentieth century.) But we surely should worry about 'human enhancement', especially if this were an opportunity available only to a wealthy elite – offering them a more fundamental form of inequality than is secured by money alone. Research on ageing exemplifies such concerns: there is an obvious incentive to pursue research with the aim of prolonging our lifespan. 'Altos' laboratories to address these problems have been set up in California (in the San

Francisco Bay area and in San Diego) and Cambridge (UK), funded by some US billionaires – people who, when young, aspired to be rich; and having achieved riches, want to be young again. Will the benefits be modest and incremental? Or is ageing a 'disease' that can be held at bay or even eliminated? Dramatic life extension, if it proved possible at all, would initially be the privilege of a fortunate few. But if it became widespread, it would plainly be a real wild card in population projections, with huge social ramifications (multi-generational families?, a later menopause?, etc.).

In addition to the distinction between curing and enhancing, many people draw a line between genetic manipulations whose effects are restricted to individuals' own bodily tissues and those that reach into the eggs or sperm and are passed down to their progeny. These manipu-lations feel vaguely Brave-New-World-ish. Yet the concept of a pure germ line (egg and sperm cells) that we alter at our peril is actually a fiction. Every parent bequeaths dozens of new mutations to their children, and research suggests the number transmitted is increased if the parent is older. The biggest threat to the integrity of the human germ line is not gene editing but middle-aged new fathers.

Manipulation of the germ line of other species also forces us to think hard about our ethical intuitions. There has, for instance, been an attempt in parts of Brazil and other areas to sterilize and thereby reduce – and even wipe out – the species of mosquito that spreads the Zika and dengue viruses; the trials recorded a 90 per cent reduction in local populations of the species. Is it bad to 'play God' in this way? Similar techniques are being proposed that could preserve the unique ecology of the Galapagos Islands by eliminating invasive species such as black rats.

If the relevant technologies continue to advance, there seems a real long-term prospect that human beings – their brains and their bodies – could be 'enhanced' via genetic and cyborg modification. Moreover, this future evolution – a kind of secular 'intelligent design' leading even to a new species – might take only centuries – in contrast to the thousands of centuries needed for Darwinian evolution.

Genetic manipulations are by no means the only ethical challenges that will confront us as biomedical science advances. We will also face acute dilemmas about treating those at the beginning and those at the end of their lives. Everyone treasures the prospect of living out more healthy years, while most people dread the prospect of being kept alive in pain or with severe disability or dementia. 'Assisted dying' or 'voluntary euthanasia' is now legalized, with safeguards, in several European countries, and several US states. In the UK public opinion is 80 per cent in favour of legalization. Professional medical opinion is shifting towards acceptance, and seems now evenly balanced; one can even find archbishops on both sides of the argument. Likewise, the ability to treat premature babies can be miraculous, but it might also mean saving children who will never flourish, laying out an ethical minefield. These are instances (and I'll raise others in this book) where there's a blurred boundary between issues where scientists deserve special attention as 'experts', and those that are ethical rather than technical, where they speak, along with the rest of us, only as citizens.

Research on viruses, such as the arch-villain of Covid-19, raises dilemmas that are both incendiary and timely. In 2011 researchers at the Erasmus University in the Netherlands and at the University of Wisconsin in the US showed it was surprisingly easy to make the H5N1 influenza virus both more virulent and more transmissible – defying the evolutionary dynamic that ordinarily trades one of these diabolical talents against the other (since a virus that kills its host can no longer use that host to spread itself). These Faustian 'gain-of-function' experiments were justified as a way to stay one step ahead of natural mutations, but they could be used with evil intent. The US federal government banned these experiments in 2014, but for reasons that seemed somewhat unclear, relaxed them three years later.[22]

Although a non-expert, I first wrote about these bio-hazards in 2003; and some better-informed colleagues thought a catastrophe was even more likely to happen than I did. On the Long Bets website,[23] maintained by the Long Now Foundation, I wagered that 'Bioterror or bioerror will lead to 1 million casualties in a single event within a six-month period starting no later than 31 December 2020.' For humanity's sake, I of course fervently hoped to lose this bet, and I was not surprised when, in 2017, psychologist Steven Pinker took me up on it, with a $400 stake (the winnings going to charity).[24] Steven had written two books documenting historical declines in violence, poverty, illiteracy and disease. He contrasted the firm data on those positive trends with the gloom of commentators whose view of the world was in his view biased by non-random samples of the worst things that happened every day, which of course tend to dominate the news media.

I responded that the positive trends, though real, can lull us into undue confidence. In the financial world, investors are all too aware that many years of gradual gains can be wiped out by a sudden loss. In biotech and pandemics (as with cyber threats – and indeed asteroid threats), the over-all risk is dominated by rare but extreme events. Moreover, as science empowers us more and more, and the world becomes increasingly interconnected, the magnitude of the worst potential catastrophes has grown unprecedentedly large.

Steven agreed that many hazards, such as wars or pandemics, fall into distributions with a 'thick tail': catastrophic events are unlikely, but not astronomically unlikely. But in a comment on the Long Bets website, he ventured that 'Moral market forces distort the odds: pessimists are seen as serious and responsible, optimists as complacent and naïve.'

Covid-19 was far more devastating than the threshold event I had wagered might occur by 2020. Both of us agree that pandemics are an ever-present threat, and probably a growing one because of more congested living and more virus-spreading air travel. The development of vaccines within a single year is surely one of our greatest scientific achievements (remember that after 40 years there is still no vaccine for the HIV virus). Covid-19 has plainly been a wake-up call that should impel us to be better prepared for future natural pandemics.

But I'd phrased my bet with Steven to exclude naturally emerging pandemics: I was envisaging an event resulting from 'bioerror or bioterror'. Do we know that Covid-19 *was* a natural pandemic? In the first week of 2021, our bet came due, and we conferred by email on how to reckon it. Our agreement was instantaneous. Though the consensus in January 2021 was that the disease was zoonotic – that is, it had jumped from animals to humans, perhaps through an intermediary host (which would make Steven the winner), we agreed the possibility of leakage of a coronavirus from China's Wuhan Institute of Virology could not be dismissed.

And we agreed to defer settling the bet until the scientific evidence was clearer.

It's a good thing we did. Subsequently the consensus unravelled and the lab-leak theory gained at least some traction. In the *Bulletin of Atomic Scientists*, the veteran science journalist Nicholas Wade prosecuted the case.[25] He reported that the Wuhan Institute was conducting gain-of-function studies; that Sars-Cov-2 bears signs of such a human-made gain; that three laboratory workers in the Institute fell mysteriously ill in autumn 2019; and that no plausible source for zoonotic transmission has yet been identified. Though most experts still think zoonosis is the more probable origin story, no open-minded scientist today could insist the case is settled. Anthony Fauci, the US infectious disease chief – who was later claimed to have been aware for years of gain-of-function experiments at the Wuhan Laboratory – added his voice to those suspecting a cover-up and seeking more openness from the Wuhan scientists. President Biden called for a report from US intelligence officials within 90 days: when they reported, they didn't rule out the 'leakage' scenario.

Realistically, given the scientific uncertainty and the deliberate opacity of Chinese authorities, our bet may never be settled. This would be frustrating (not so much for us as for science and public health), but it is not without some comfort. Proof of a lab leak could, as the legal scholar Stephen Carter noted,[26] 'give the coronavirus saga what it's lacked: a villain' and 'the formless fear that has immobilized most of the world for the last year and a half, at last given a target, might coalesce into fury'. It could also turn people against science and invite crippling, over-broad regulation, slowing progress against disease, death and disability. Whatever the origin of Covid-19, we can't rule out lab leakage in the future (recalling, for example, that a serious foot-and-mouth outbreak in the UK was caused by a leakage from the Pirbright laboratory in Surrey in 2007). There is surely a case for enhancing security and independent monitoring of the 'level 4' laboratories around the world researching lethal pathogens.

But can we rule out a future release that could be intentional, rather than accidental? To be sure, governments, and even terrorist groups with specific aims, will always be inhibited from releasing bioweapons, because no one can predict where and how far they can spread. The real nightmare would be a deranged loner with biotech expertise who believed, say, that there were too many humans on the planet and didn't care who became infected, or how many. The ultimate bioweapon would combine high lethality, the transmissibility of the common cold (or the Omicron variant of Covid-19) and a long asymptomatic period allowing massive spreading before countermeasures could be taken.

The odds militate against this worst-case scenario. In the real world, technological feats are rarely pulled off by a single evil genius, and complex plots tend to be pre-empted by surveillance, or derailed by mishaps or incompetence. But even one instance is one too many if its consequences cascade globally.

In the early days of recombinant DNA research, back in 1975, the world's leading molecular biologists met in Asilomar, California, and drew up guidelines on which kinds of experiments should not be done. This was an encouraging precedent, and it has been followed by similar meetings, convened by national academies, journal editors and government officials, to discuss recent biotechnologies in a similar spirit of proactive caution. But now, nearly 50 years after the first Asilomar meeting, the researchers are no longer a small academic community concentrated in Europe and North America; expertise has hugely expanded and spans the globe. Moreover, much of the research is done in commercial companies rather than openly in academia. The dangers loom ever larger. Regulation of biotech is needed even more today – hopefully in a concerted universal fashion, following the advice from professional groups, academies and bodies like the WHO.

But I think there's a serious worry about effective global enforcement of whatever regulations are imposed. Could they be enforced throughout the world any more effectively than the drug laws can, or the tax laws? Whatever can be done may be done by someone, somewhere. That is the stuff of nightmares. In contrast to the elaborate, conspicuous special-purpose equipment needed to create a nuclear weapon, which can feasibly be monitored by international inspectors, biotech involves small-scale, dual-use technology that will become widely accessible. An increasing number of individuals will acquire the requisite expertise – indeed, biohacking is burgeoning as a hobby and competitive game. And there are many hundreds of laboratories around the world in which dangerous pathogens are being studied and modified.

We're kidding ourselves if we think that those with technical expertise will all be balanced and rational: expertise can be allied with fanaticism – not just the types of fundamentalism with which we are currently familiar, but that exemplified by some New Age or 'conspiracy' cults, extreme eco-freaks who think there are too many humans on this planet, and the like. And there will be individuals, with the mindset of those who now unleash computer viruses – the mindset of arsonists. The global village will have its village idiots. In a future era of vast individual empowerment, where even one malign or foolish act could be one too many, how can our open society be safeguarded? The rising empowerment of tech-savvy groups (or even individuals), by biotech – and by cybertech as well – will pose an intractable challenge to governments, and aggravate the tension between three values we cherish: freedom, privacy and security. And that's a tension that will be balanced very differently in the US and in China.

The world was under-prepared for Covid-19, however it originated. More broadly, it is unprepared for the intellectual, moral and practical challenges posed by burgeoning biotechnology. These challenges call for clear thinking and well-crafted policies that recognize both its stupendous potential boon to human flourishing and its stupendous potential risk to human safety – indeed to humanity itself. It's an arena where the regulations that ethics and prudence require may not only prove controversial but near impossible to enforce. And I fear the same is true for another transformative technology: robotics and artificial intelligence (AI).

1.4 Computers, robots and AI

'AlphaGo Zero', programmed by computer scientists at Deep Mind, a London-based company now owned by Google/Alphabet, famously beat human champions in the games of Go and Chess. It was given just the rules, and 'trained' by playing against itself over and over again, for just a few hours. This signalled an expansion in the range of tasks where machines can outsmart even the brightest of us.

AI can cope better than humans with complex fast-changing networks – traffic flow, or electric grids – and process immense datasets. The Chinese could have an efficient planned economy that Marx or Stalin could only dream of. And it can help science too – in drug development, and by accelerating and surpassing unaided human efforts to determine the structure of protein molecules. It can perhaps even master ten-dimensional geometry well enough to settle one of the biggest mysteries in my academic field: whether string theory can really describe our universe.

The societal impacts of machine learning and AI are already mixed. Systems and networks will become more intrusive and pervasive. Records of all our movements, our health and our financial transactions will be in the 'cloud', managed by a multinational quasi-monopoly. The incipient shifts in the nature of work have been addressed in several excellent books by economists and social scientists.[27] The workers most likely to be displaced are those doing mind-numbing work in, for instance, call centres or warehouses. If these people could find alternative jobs where 'being human' is the prime qualification – such as carers for the young, old and sick – this would be a 'win–win' transition. But bringing it about would require taxation of the companies, and hypothecating the funds so raised for the support of far greater numbers of carers; they deserve greater security and standing than the market (or a government motivated by austerity and tax cutting) offers them today.

Machines are trained by speedily absorbing datasets. They will therefore mimic any biases inherent in that data. That's why it's concerning that AI is being deployed, for instance, to draw up a shortlist from fields of job applicants. Even more scarily, some companies use 'facial recognition' software to analyse videos of applicants, to allegedly infer from their expressions their emotions and character. If we're denied a job – or sentenced to imprisonment, recommended for surgery or even denied credit by our bank – we would expect to be given the reasons, and have a chance to contest them. If such decisions are delegated to algorithms, we are entitled to feel uneasy, even if presented with evidence that, on average, the machines make better decisions than the humans they have usurped. (One must accept, however, that their judgements may sometimes be more consistent – there have been claims[28] that some human judges impose, on average, different sentences on criminals before and after the lunch break!)

Ethical tensions are already emerging when AI moves from the research phase to being a potential money-spinner for global companies. For instance, when Google/Alphabet took over Deep Mind, an ethical committee was disbanded that had been set up to address privacy issues for medical data held by the UK's National Health Service.

It is of course the *speed* of computers that allows them to learn rapidly when big 'training sets' are available for them to 'crunch'. But learning about human behaviour – acquiring 'common sense' – won't be so easy for them. It involves observing actual people in real homes or workplaces. A machine would be sensorily deprived by the slowness of real life – it would be like watching trees grow is for us. But sensor technology that allows robots to interact more sensitively with the real world is advancing fast. Boston Dynamics has built a robot, ATLAS, which can tackle an obstacle course and perform gymnastics – better than most humans, if not at Simone Biles' level.

But there may be trouble ahead. Futuristic books by philosopher Nick Bostrom[29] and physicist Max Tegmark[30] portray a 'dark side' – where AI 'gets out of its box', infiltrates the internet of things or the global financial system, and pursues goals misaligned with human interest – or even treats humans as an encumbrance. Some AI pundits take this seriously, and think the field already needs guidelines – just as biotech does – to ensure 'responsible innovation'. But others (like the inventor of the Baxter robots, Rodney Brooks) regard these concerns as premature – and think it will be a long time before artificial intelligence need worry us more than real stupidity.

Be that as it may, it's likely that society will be transformed by autonomous robots, even though the jury's out on whether they'll forever be 'idiot savants' or one day display superhuman capabilities – whether we should worry more about breakdowns and bugs, or about being outsmarted. Predictions of how fast AI will advance range from those who predict runaway progress to a 'singularity' (when technological advance will slip out of our hands, becoming irreversible and unstoppable) within 30 years, and those who doubt this will even happen.

And this leads to a digression.

It's always harder to forecast the speed of technological advances, and their rate of adoption, than their direction. Sometimes there's a spell of exponential progress – like the global spread of IT and smartphones in the last two decades. But in the longer term, progress in any area is often better described by a graph that rises steeply and then levels off, either because of fundamental barriers, or (more often) because incentives and demand ease off because of a more alluring rival innovation. Here are two historical examples from the aerospace sector.

From Alcock and Brown's first transatlantic flight in 1919 to commercial flights of the first jumbo jet, Boeing's 747, took 50 years. But 50 years later we still have the jumbo jet – and during that half-century Concorde came and went! Even if supersonic commercial flight returns it won't be until the 2030s. And – a second example – only 12 years elapsed between Sputnik 1, in 1957, and the Apollo moon landings: but 50 years later that's still the high point of human spaceflight.

I'd speculate about two future trends that may replicate this pattern. Experts are getting less optimistic about how quickly we'll have stage 5 driverless cars – vehicles programmed so that they cannot merely cope with motorways, but allow a human 'driver' to relax in the back seat even in London traffic and on winding country roads with stray animals, etc. And smartphones may by now be as capable and complicated as most users want, so the iPhone 20 may not be too different from the iPhone 13.

Of course, overall technological innovation can continue – driven by a succession of surges, each experiencing a phase of exponential growth followed by 'saturation'. For instance, smartphone technology may 'saturate' due to being trumped by consumer demand for virtual reality (the 'metaverse'), holograms and suchlike.

But it's in my own special field, space exploration, that robots and AI have the greatest scope and raise fewest concerns. During this century, the entire solar system will be explored by flotillas of tiny robotic craft. But will people follow them? The practical case is getting weaker as robots acquire better sensors and more 'intelligence'. More than a decade ago, NASA sent a vehicle called 'Curiosity' to Mars. It trundled very slowly across a giant crater: if it sensed a rock in its path, it had to 'report back' to its handlers and get instructions on how to change course. A later vehicle, 'Perseverance', which landed on Mars in 2021, had sufficient intelligence to work out a safe rock-avoiding course for itself. And a decade or two from now, future probes may have the abilities to identify unusual geological formations as reliably as a human geologist could.

If humans venture far from the Earth – at vastly more expense than sending a robot on the same voyage – it will primarily be as an adventure. Some will go, in this spirit, to the Moon and perhaps a few to Mars – but I think they should be privately sponsored and prepared to embark on cut-price missions far riskier than a Western government could impose on publicly funded civilians. (And a return ticket to Mars would cost far more than twice a one-way ticket because of the equipment that would need to be brought from Earth to enable a well-provisioned craft to be launched from the Martian surface for the return journey.) But don't ever expect mass emigration from Earth. It's a delusion, and indeed a dangerous one, to think that space offers an escape from Earth's problems. Coping with climate change is a doddle compared to terraforming Mars. There's no 'Planet B' for ordinary risk-averse people. We must cherish our Earthly home.

Indeed, a cosmic perspective actually strengthens our concerns about what happens here and now, because it offers a vision of just how special our Earth is, and how prolonged and prodigious life's future could be. In the aeons that lie ahead, even more marvellous diversity could emerge: the unfolding of intelligence and complexity could still be far from its culmination. We're all surely mindful of the heritage we've inherited from our forebears. If our generation are negligent stewards, we shall not only jeopardize the welfare of our children and grandchildren but risk foreclosing vast future potentialities.

1.5 Avoiding doom

The global challenges I've addressed all require us to plan up to a century ahead – we should care about the life chances of those just born who can expect to survive into the twenty-second century, especially in the context of climate and biodiversity. But this brings us to a question that has been discussed by philosophers like Frank Ramsay and Derek Parfit: our obligations to 'possible people' who aren't yet born.[31] We plainly need to minimize the probability of global catastrophes, but how much worse is a catastrophe if it not only truncates the lives of those now living, but also forecloses the existence of future generations?

The historical record reveals that 'civilizations' have crumbled and even been extinguished. Famously, Jared Diamond discussed such events in his book *Collapse*;[32] environmental policy analyst Luke Kemp and others are now analysing evidence for larger numbers of such events. But there's a crucial difference today: our world is so interconnected that a catastrophe couldn't hit any region without its consequences cascading globally. The worldwide spread of Covid-19 makes us all too aware of this: it's a portent for even worse that could be yet to come.

Pandemics not only spread faster – and spread more globally – than they did in the past: they cause far worse societal breakdown. In the fourteenth century, European villages continued to function even when bubonic plague (the Black Death) halved their populations. In contrast, we're all too aware today that societies are vulnerable to serious unrest as soon as hospitals are overwhelmed. And this could occur before the fatalities rose much above 1 per cent. The mortality figure for the US and UK due to Covid-19 is about 0.4 per cent. It has been three times higher in Peru, and probably in other countries where the records are less likely to be complete and where vaccination rates are still low. There's likewise huge societal risk – perhaps even of reversion to anarchy – from cyber attacks on infrastructure: we surely wouldn't have coped so well with lockdown during the pandemic had we not felt able to rely so confidently on the availability and robustness of the internet. What would have happened if the electricity grid or the internet had failed in the middle of the Covid-19 pandemic?

It needs little imagination to contemplate a collapse – societal or ecological – that would be a truly global setback. The setback could be temporary. On the other hand, it could cascade and spread so devastatingly (and entail so much environmental or genetic degradation) that the survivors would take centuries to regenerate a civilization at the present level.

Biologists should avoid the creation of potentially devastating genetically modified pathogens, or large-scale modification of the human germ line. Cyber experts furthering the beneficent uses of advanced AI should avoid scenarios where there seems even a minuscule chance of a machine 'taking over'. Many people are inclined to dismiss such scenarios as too unlikely to deserve our attention except as science fiction. Moreover, they may validly argue that innovations have an upside too and indeed are crucial for humanity's future. Application of the 'precautionary principle' has an opportunity cost— in Freeman Dyson's phrase, there is a 'hidden cost of saying no'.

It is nonetheless worth considering such extreme scenarios as a thought experiment – and to ruminate on the ethical issues they entail. We can't rule out human-induced threats far worse than those on our current risk register. Indeed, we have zero grounds for confidence that human civilization – or even humanity itself – can survive the worst that future technologies could bring. The 'existential' threats, with consequences that cascade through many generations, are a mega-version of those that arise in climate policy, where it's controversial how much weight we should give to those who will live a century from now. We indeed need to think longer term than we do.

But, to return to the question of our obligations to the more remote far-distant future, human extinction could foreclose the existence of billions, even trillions, of future beings — and indeed perhaps of an open-ended posthuman diaspora far beyond the Earth. Many ethicists would claim that the welfare of 'possible people' matters as much as that of the present generation. But even if one accepts this, it's not obvious how strongly the remote future should be weighted in present-day decisions. This is because we have decreasing confidence about how present actions will resonate deeper into the future. Moreover, even though human character has probably changed little for many

Chapter 2

Meet the Scientists

2.1 Science and culture: past and present

In the 1660s the founder-members of the Royal Society – Christopher Wren, Robert Hooke, Samuel Pepys, and other 'ingenious and curious gentlemen' (as they described themselves) – met regularly. Their motto was to accept nothing on authority. They did experiments; they peered through newly invented telescopes and microscopes; they dissected weird animals. One experiment involved transfusion of blood from a sheep to a man (who survived). But, as well as indulging their curiosity, they were immersed in the practical agenda of their era: improving navigation, exploring the New World, and rebuilding London after the Great Fire. Some of them were religious, but their inspiration was Francis Bacon, for whom there were two goals for scientists: to be 'merchants of light', and to promote 'the relief of man's estate'. A century later (1769) the 'American Philosophical Society' was founded in Philadelphia for the

'Promotion of Useful Knowledge' with Benjamin Franklin as its first President.

In the late eighteenth and early nineteenth centuries, the same interdisciplinary spirit prevailed. Richard Holmes's fascinating book *The Age of Wonder* describes the affinity between the sciences – especially the fruits of exploration by Captain Cook, Joseph Banks and others – and the creativity of poets like Coleridge and Shelley. There was no split between 'two cultures', but a boisterous intermingling of scientists, literati and explorers. The Royal Society (where Joseph Banks was president for 42 years, and had sufficient wealth to subsidize scientific work) was then supplemented in the UK by the Royal Institution (RI) as a focus of such interactions.

The RI was bankrolled by the hyper-talented but roguish adventurer Count Rumford, who donated sufficient funds to provide a fine building in Albemarle Street in central London. Rumford made important discoveries about the nature of heat, as well as being a fertile inventor. After supporting the losing side in the war of American Independence (and marrying a wealthy American), he moved to England, and then to Bavaria where he made many of his inventions, re-organized the army, and acquired the title 'Count of the Holy Roman Empire' from the Elector of Saxony. He later lived in Paris, marrying the widow of Lavoisier, the French scientist who discovered the role of oxygen in combustion. Rumford envisaged the RI's mission as research, but also as the dissemination of scientific understanding among the wider population. The RI was fortunate in the calibre of its first two directors, Humphrey Davy and Michael Faraday; both were exceptional scientists, but also promoted 'outreach', mainly via weekly 'discourses', involving lectures which attracted a London elite – and which still continue today, albeit with less allure.

The nineteenth century saw a surge of public enthusiasm for science: this was the era when specialized learned societies were founded in the UK, along with the British Association for the Advancement of Science (BA), to bring science to the public. The BA, founded in 1831, had a wider social and geographical range than the RI. In 1837 when it met in Newcastle in the north of England, it is recorded that the geologist Adam Sedgwick gave an open-air lecture to 3,000 'colliers and rabble'. And it is testimony to public enthusiasm that the 1851 Great Exhibition, showcasing the achievements of technology in the specially constructed Crystal Palace, received 6 million visits. The US counterpart of the BA, the American Association for the Advancement of Science, was founded in 1848; the magazine *Scientific American* dates from 1845.

We are reminded by historians that the technologies that triggered the industrial revolution developed independently of what we'd now call science (and was then called natural philosophy). Indeed, academic science wasn't part of the student curriculum until surprisingly late. The Cambridge academic William Whewell, a genuinely impressive polymath in many ways (and the first to coin the word 'scientist'), thought all students should learn the 'eternal truths' of mathematics – and of theology! He deemed scientific ideas too uncertain and transitory to be taught formally, though he conceded that anything that had been known for at least 100 years was probably OK. And it's often forgotten, in more recent discussion of the balance between 'humane' and 'technical' education, that the traditional university curriculum across Europe was largely vocational – preparing students for medicine, the law and the church – and that mathematics was widely taught. The numbers attending universities in earlier centuries were of course minuscule compared to today. Post-18 education is becoming the norm, but its format and institutional underpinnings aren't yet appropriate for the twenty-first century (which we will turn to in chapter 4).

Attitudes to science shifted in the second half of the nineteenth century. Cambridge's Cavendish Laboratory was founded by a munificent gift – and headed successively by three of the greatest physicists of that period – James Clark Maxwell, Lord Rayleigh and J. J. Thomson – whose discoveries formed the basis for much of twenty-first-century technology.

However, it was a biologist, Charles Darwin, whose impact on Victorian thought was most profound – and resonates even more today. His concept of natural selection has been described by the philosopher Daniel Dennett, with only slight hyperbole, as 'the best idea anyone ever had'. His insights are pivotal to our understanding of all life on Earth, and the vulnerability of our environment to human actions.

At the end of the nineteenth century the Scottish physicist Lord Kelvin – who gained wealth by his contributions to laying cables across the Atlantic for telecommunication – famously claimed that physics was essentially 'done': all that was left was to refine the values of a few 'physical constants' and fill in some details. But this assertion didn't connote complacency or triumphalism. On the contrary, Kelvin had an over-modest and constricted view of the scope of science: he never conceived that we'd one day interpret the properties of materials – and even living things – in terms of their molecular structure; nor that physicists would probe deep into the nuclei of atoms (revealing an unenvisioned new source of energy); nor develop new insights into gravity, space and time. But in the twentieth century, of course, all these breakthroughs happened. The scope and scale of science widened hugely – but the consequent greater specialization has balkanized the map of learning, and created damaging gulfs between specialists in different fields, and with the wider public, even as science impacts more deeply on everyone's lives.

Our conceptual horizons have expanded hugely; no new continents remain to be discovered. Our Earth no longer offers an open frontier, but seems constricted and crowded – a 'pale blue dot' in the immense cosmos. Today's scientists are specialists, not polymaths. But despite all that has changed, the values and attitudes of the Royal Society's founders are enduring ones; though nowadays too few engage broadly with society and with public affairs.

But such engagement is needed more than ever before. Now that the spin-offs from science are so pervasive in our lives – and will be crucial in tackling the great global threats described in chapter 1 – everyone should have a voice in ensuring that it is applied ethically, and to the benefit of both the developing and developed world. We must confront the widely held anxieties that genetics and artificial intelligence may 'run away' too fast. As citizens, we all need a feel for how much confidence can be placed in science's claims.

Quite apart from its impact on our lives, it is surely also a cultural deprivation not to appreciate the panorama offered by modern cosmology and Darwinian evolution – the chain of emergent complexity leading from some still-mysterious beginning to atoms, stars and planets – and how, on our

planet, life emerged, and evolved into a biosphere contain-
ing creatures with brains able to ponder the wonder of it all.
This common understanding should transcend all national
differences – and all faiths too. Moreover, scientific discoveries and their applications are so crucial to our collective
future that all citizens need to appreciate the scope and
limits of what science can offer. The rest of this chapter
expands on this theme.

2.2 What scientists do

Science is a truly global culture. Its universality is especially compelling in my own subject of astronomy. The dark
night sky is an inheritance we've shared with all humanity,
throughout history. All have gazed up in wonder at the
same 'vault of heaven' but interpreted it in diverse ways.
There is a natural fascination with the big questions: was
there a beginning? How did life emerge? Is there life in
space? And so forth.

The simplest building blocks of our world – atoms –
behave in ways physicists can understand and calculate.
And the laws and forces governing them are universal:
atoms behave the same way everywhere on Earth – indeed,
spectroscopy reveals that they are the same even in the
remotest stars. These 'basics' are firmly enough understood
to enable engineers to design all the artefacts crucial to our
modern world, from silicon chips to rockets.

Our environment is far too complicated for every detail to
be explained, but our perspective has been transformed by
great insights – great unifying ideas. For instance, the concept of continental drift allows geophysicists to interrelate

geological and ecological patterns across the globe. And Darwin's great insight revealed the overarching unity of the entire web of life on our planet.

The more we understand the world, the less bewildering (but more amazing) it becomes, and the more we're able to change it. Nature displays many patterns. There are even patterns in how we humans behave: how cities grow, how epidemics spread, and how technologies develop. Computers have had an especially important impact here. They're essential in the analysing of data, but have also been important for 'modelling' systems where we can't do actual experiments: weather and climate are obvious and important examples. But those studying astronomy (my own field) have perhaps gained most. We plainly can't do any experiments on our subject matter: we can only observe. However, we can now experiment on a speeded-up 'virtual universe' on our computer: crashing another planet into the Earth to see if that's how our Moon might have formed; computing how a massive star explodes as a supernova. We can simulate what would happen if two galaxies collide – when, for instance, the Andromeda Galaxy crashes into our Milky Way in 4 billion years. A computer can in principle simulate how our entire visible universe, starting as a hot dense amorphous 'fireball', expanded and cooled, transforming itself nearly 14 billion years later into the complex cosmos we see around us, and of which we are a tiny part.

There are occasional mega-breakthroughs that deserve to be called 'scientific revolutions'.[1] But most advances aren't revolutionary: instead, they corroborate, transcend or generalize the concepts that went before, or (most often) just fill in some details. For instance, Einstein didn't 'overthrow' Newton's theory of gravity. He devised a theory that had broader scope and gave deeper insights into the nature of space and gravity, but Newton's laws are still good enough to predict the trajectories of spacecraft. (There is, though, one practical context where Einstein's refinements are needed: the accuracy of the Global Positioning Satellites (GPS) used in satnav systems would be fatally degraded if proper allowance wasn't made for the slight difference that relativity theory predicts between the rates at which clocks tick on Earth and in orbit.)

Scientists are tough critics of new ideas. They have a professional incentive to uncover errors. That's because the greatest esteem goes to those who contribute something unexpected and original, and especially to those who can overturn a consensus. That's how initially tentative ideas get firmed up – not only on topical issues like climate change and the control of pandemics, but (to cite some examples from earlier years) the link between smoking and lung cancer, and between HIV and AIDS. But that's also how seductive theories get destroyed by harsh facts. Science is 'organized scepticism'.

Exciting new scientific claims are scrutinized far more widely and thoroughly than in the past. That speeds up progress in a subject like mine; but it is of course much more practically important that the same can be done with medical data from around the world – a fact that's yielded crucial benefits in coping with pandemics by allow-ing more rapid identification of new variants of pathogens, and quicker testing and validation of drugs and vaccines. But as I'll discuss later, there are downsides: the com-mercial, political and careerist pressures of 'mass science' can generate unhealthy competition and even fraud. And bombardment by confusing messages on social media makes it hard for the public to distil a balanced view of controversies.

As a Cambridge student in the 1960s, I watched at close hand a stand-off between Fred Hoyle and Martin Ryle – two outstanding scientists utterly different in their personal and professional styles – on whether the universe had emerged from a 'Big Bang' (as Ryle argued) or whether, as Hoyle contended, it had existed for ever in a so-called 'steady state'. New evidence in the mid-1960s settled this debate in Ryle's favour – an outcome to which Hoyle was never fully reconciled, though by the end of his life he was advocating a compromise 'steady bang' theory that gained little traction with others. Hoyle enjoyed robust controversy. Ryle shunned it. But, to be fair to Ryle, an experimenter who spends years designing and constructing an instrument (and he was very much a 'hands-on' instrument builder) understandably develops an exaggerated perception of its importance. In contrast, theorists can be relaxed about jettisoning theories: those with fertile minds can quickly devise new ones.

There have likewise been high-profile vendettas on conceptually important issues in evolutionary biology and sociobiology. But fewer scientific disputes today become so deeply personalized. This doesn't reflect the sweeter dispositions of a younger generation of scientists – it's because, as data accumulate, there is progressively less scope for viable but strongly divergent hypotheses; and there is a growing incentive towards collaborative rather than isolated research. But serious nastiness can be generated when there's duplication, and competition for priority, in overcrowded 'fashionable' fields. (I feel fortunate to have worked on topics where the ratio of problems to people is gratifyingly high – so any modest new insight I can achieve doesn't advance the subject by a mere few months before someone else would have done it anyway.)

When rival theories are in contention there is eventually just one winner – at most. Sometimes, a crucial piece of evidence clinches the case. This happened for the Big Bang cosmology when radio engineers detected weak background microwaves, best interpreted as a relic of the cosmos's hot dense beginnings. Likewise, continental drift was only accepted when direct evidence was found of 'sea-floor spreading' – an upwelling of material along vents in the ocean bed which push apart the continents, as though they were plates floating on some substratum. (An eminent geophysicist of an older generation, Harold Jeffreys, resolutely opposed continental drift, believing that the Earth's interior was too rigid for this flow to occur – and although he made the wrong judgement in this case, it is appropriate to insist on a higher standard of proof before accepting an idea that seems implausible.) But in these two instances only a few experts 'held out'. In other cases, an idea gains only a gradual ascendancy; alternative views get marginalized until their leading proponents die off. Or the subject moves on, and what once seemed an epochal issue is bypassed or sidelined.

Sometimes the balance of opinion is biased by ideology, leading to the attempted suppression of ideas that eventually triumph. The most infamous example was that of Galileo Galilei in the seventeenth century. In his trial by the Inquisition, he was found 'vehemently suspected of heresy'. The great poet John Milton visited Galileo in 1638 during his 'grand tour'. In his impassioned plea for freedom of speech, 'Areopagitica', published in 1642, he lamented that he 'found and visited the famous Galileo grown old, a prisoner to the Inquisition, for thinking in Astronomy otherwise than the Franciscan and Dominican licensers thought.'

There have been other such cases of suppression over the centuries, some consigning scientists to a fate far harsher than Galileo's. A famously tragic episode involved geneticists working in the Soviet Union in the Stalin era. In the 1920s Nikolai Vavilov founded and directed the Lenin All-Union Academy of Agricultural Sciences; it researched successfully on techniques to improve plant and animal breeding. However, the maverick geneticist Trofim Lysenko gained Stalin's support and usurped Vavilov's role in 1935. He believed that acquired traits are inherited, claimed that heredity can be changed by 'educating' plants, and denied the existence of genes. In subsequent purges, several 'mainstream' geneticists were shot. Vavilov himself was arrested in 1940 and died of starvation in prison three years later. Lysenko's influence lingered, as a 'drag' on Soviet biology, until the 1960s.

In the present era, we are familiar with those who can be broadly described as 'deniers' – people who react with extreme scepticism to what the majority regards as strongly supported evidential claims: the health risks of tobacco, the benefits of vaccination, the dangers of unchecked climate change, and so forth. Their bias may be ideological; or it may (as in tobacco or fossil fuel companies) reflect a straight commercial judgement that it's in their interests to spread doubt and uncertainty about scientific findings. In either case it can distort both media coverage and funding – and, more seriously, impede adoption of important programmes necessary for safety or human welfare.

If you ask scientists themselves what they are working on, you will seldom get an inspirational reply like 'seeking to cure cancer' or 'understanding the universe'. Rather, they will focus on a tiny piece of the puzzle and tackle something that seems tractable. Scientists are not thereby ducking the big problems, but judging instead that an oblique approach can often pay off best. A frontal attack on a 'grand challenge' may, in fact, be premature. To take a historic example: 50 years ago President Richard Nixon declared a 'war on cancer', envisaging this as a national goal modelled on the then-recent Apollo Moon-landing programme. But there was a crucial difference. The science underpinning Apollo – rocketry and celestial mechanics – was already understood, so that, when funds gushed at NASA, the Moon landings became reality. But, in the case of cancer, the scientists knew too little to be able to target their efforts effectively.

It is amusing to think of other examples where a head-on approach would surely have failed. Suppose that nineteenth-century innovators had wanted to develop better machines to reproduce music. They could have made very elaborate mechanical organs or pianolas, but these efforts wouldn't have accelerated the advent of the CD player or Spotify. And a medical programme seeking ways to see through flesh wouldn't have stimulated Roentgen's serendipitous discovery of X-rays.

Nixon's cancer programme, incidentally, facilitated a lot of good research into genetics and the structure of cells. Indeed, the overall research investment made in the twentieth century has paid off abundantly. But the pay-off happens unpredictably, and after a time lag that may be decades long. This is why much of science is best funded as a public good. A fine exemplar of this point is the laser, invented in

1960. It exploited basic ideas that Einstein had developed more than 40 years earlier, but its inventors couldn't have foreseen that lasers would later be used in, for instance, eye surgery and in DVD players.

A 'war on cancer' can now have a better focus than in Nixon's era: far more is known about cell biology. And there have been spectacular advances in epidemiology and our understanding of viruses – as the speedy development of vaccines against Covid-19 has shown. But today there are new 'grand challenges' where basic understanding languishes at the level of cancer research in the Nixon era – brain research, for instance. Programmes in the US and the EU for a 'moonshot' programme in brain science have generated scepticism and controversy.[2] We don't design aircraft with flapping wings to emulate birds; likewise, the differences between how computers operate and how brains 'think' may be more fundamental, and more intractable, than generally presumed, and still too mysterious to merit such massive coordinated programmes.

The word 'science' is being used here, it should be said, in a broad sense, to encompass technology and engineering; this is not just to save words, but because all these disciplines are symbiotically linked (in ways they weren't before the mid-nineteenth century). The mental process of problem solving motivates us all – whether one is an astronomer probing the remote cosmos, or an engineer facing a down-to-earth design conundrum. The patterns or 'laws' we have discerned as part of this problem solving are the great triumphs of science. To discover them required dedicated talent – even genius – or amazing good luck. But to grasp their essence isn't so difficult: most of us appreciate music even if we can't compose or even perform it. Likewise, the key ideas of science can be accessed and enjoyed by almost everyone; the technicalities may be daunting, but these are less important for most of us, and can be left to the specialists.

2.3 Communication and debate

The computer and IT revolution has inordinately widened and speeded up the exchange and communication of information; it has also reduced the advantage you have if you work in a major institution. I have spent most of my working life in one of the world's leading centres for astronomy; we have a large international research group. However, much of our research entails collaboration with remote partners. Our working day often consists mainly of being video-linked or exchanging texts with collaborators in remote places, and analysing data transmitted from around the world. This has speeded up the subject, but it has other benefits: our colleagues in, for example, South Africa (where there's a vibrant scientific community), who in the pre-internet era depended on frustratingly slow and inefficient mail services, now aren't really handicapped as they can get the data as quickly as we do. And there's an important benefit for researchers studying transient phenomena. When an observatory reports a sudden cosmic event, observers worldwide can be alerted instantly so they can make complementary or follow-up observations. For instance, when a remarkable new phenomenon was detected – a pulse of 'gravitational waves', lasting just a few seconds, generated by a merger of two neutron stars – follow-up data came from 70 observatories, covering all wavebands, and were published in a single paper with 1,000 authors.

Information technology (IT) has democratized science; information and data can be accessed rapidly from anywhere. To take another example from my own subject, the plates from an important photographic sky survey made by a telescope at the Palomar Observatory are stored in some archive in California; only a few copies were made and distributed to major centres. The survey (like all recent ones) is now digitized. Just as scholars don't now need to go to a library to read a manuscript, so scientists in many fields can access or download any body of data and analyse it. (There are of course some constraints. For instance, projects in human genetics or epidemiology, using data from individuals, may need to be anonymized.)

Traditionally, scientific findings reached the community's attention only after being published in an academic journal. This custom actually dates back to the seventeenth century. In the 1660s, the Royal Society started to publish *Philosophical Transactions*; this was the first scientific journal, and continues to this day. Authors were enjoined to 'reject all amplifications, digressions and swellings of style'. This journal pioneered what is still the accepted procedure of 'peer review' whereby ideas are criticized, refined and codified into 'public knowledge'. It aimed at international coverage from the start: its founder-editor was a Dutchman, Henry Oldenberg and the title page quaintly described its aim as 'giving some account of the present undertakings, studies and labours of the ingenious in many considerable parts of the world'. Over the centuries, it published Isaac Newton's researches on light, Benjamin Franklin's experiments on lightning, reports of Captain Cook's expeditions,

Volta's first battery – as well as, up to the present day, many of the triumphs of modern research.

Much has changed, however. It's now routine in many subjects for new research papers to be posted on a website before being published in a journal. Every morning, when still at home, I look at the abstracts of astronomical papers posted overnight on a website. Any especially noteworthy or controversial paper that has been posted will be a topic of coffee-time conversations that day in institutes around the world – and of social media discussions too.

The peer review procedure for quality control is, however, under increasing strain, due to competitive or commercial pressures, 24-hour media, and the greater scale and diversity of a scientific enterprise that is more widely international. (And, of course, scientific journals are themselves now mainly distributed electronically rather than as paper copies.) In some Chinese universities, faculty members got a substantial bonus if they published in a 'high profile' international journal. Partly in response to such incentives, there's now a proliferation of 'journals' that purport to be serious and peer-reviewed, but which are actually commercial scams, willing to publish anything for a fee. And there are, increasingly, pressures to bypass the frustratingly slow procedures traditional in publishing printed journals.

A celebrated (but deplorable) departure from traditional norms happened back in 1989 when Stanley Pons and Martin Fleischmann, then at the University of Utah, claimed at a press conference to have generated nuclear power using a tabletop apparatus. If credible, this 'cold fusion' fully merited the hype it aroused: it would have ranked as one of the most momentous breakthroughs since the discovery of fire.

But doubts set in. Extraordinary claims demand extraordinary evidence, and in this case the evidence proved far from robust. Inconsistencies were discerned; and others failed to replicate what Pons and Fleischmann claimed to have done. Within a year, there was a consensus that the results had been misinterpreted, though even today there are still some 'believers'.

The 'cold fusion' claims bypassed the traditional quality controls of the scientific profession, but did no great harm in the long run, except to the reputations of Pons and Fleischmann. Indeed, in any similar episode today, exchanges via the internet would have led to a consensus verdict even more quickly. Whenever a paper attracts wide interest, the collective informal scrutiny it receives can be more rigorous than any formal refereeing.

But this fiasco holds an important lesson. What's crucial in sifting error and validating scientific claims is open discussion. If Pons and Fleischmann had worked not in a university but in a lab whose mission was military, or commercially confidential, what could have happened? If those in charge had been convinced that their scientists had stumbled on something stupendous, a massive programme might have got under way, shielded from public scrutiny and wasting huge resources. (Indeed, just such waste has occurred, more than occasionally, in military laboratories. One example was the 'X-ray laser' project spearheaded by Edward Teller and Lowell Wood at the Livermore Laboratory as part of President Reagan's 'Star Wars' initiative in the 1980s.)

Of course in the commercial world the pressure to hype dubious or even bogus claims is enormous. For instance, a company called Theranos achieved, in 2015, a peak capitalization of US$9 billion by claiming to have built a small gadget incorporating computer chips that could quickly analyse a tiny blood sample and diagnose an individual's 'disease profile'. This company gained credibility partly because its personable young founder Elizabeth Holmes (who hadn't even completed a doctoral degree) had persuaded a group of US political luminaries, none with any relevant expertise, to lend prestige and credibility to her Board. The huge hype garnered investments from wealthy celebrities; its new building was opened by then-Vice-President Joe Biden. But things soon fell apart; a staff member shared his concerns with a *Wall Street Journal* reporter, whose reports triggered plummeting of the share price, accompanied by resignations and acrimonious lawsuits. In January 2022, Holmes was convicted of fraud.

The imperative to foster openness and debate is a common thread through all the examples I've discussed. It ensures that any conclusions that emerge are robust, and that science is 'self-correcting'. Even wider discussion is needed when what's in contention is not the science itself, nor its commercial value, but whether its applications are ethical and safe. When important principles are involved, such discussions should be open to all of us, as citizens – and, of course, should engage our elected representatives. Sometimes this has happened, and constructively too. In the UK, ongoing dialogue with parliamentarians led, despite divergent ethical stances, to a generally admired legal framework for regulating research on embryos. A committee chaired by the philosopher Mary Warnock proposed that experiments should be allowed only on embryos less than 14 days old; this set a standard that many other countries followed. The UK also formulated widely accepted guidelines on stem cells – a contrast to what happened in the US where experiments were prohibited if federally funded, but allowed if the funding came from a state or a private foundation.

But the UK had failures too: the GM crop debate was left too late, to a time when opinion was already polarized between eco-campaigners on the one side and commercial interests (especially the Monsanto company) on the other. Indeed, perhaps one actual benefit of Brexit is that the UK won't be bound by the over-restrictive EU regulations on GM crops. The EU ban extends to the use of CRISPR–Cas 9 gene-editing, which is actually less objectionable as it is narrowly targeted and entails no 'trans-species' transfer of genes.

In debates on bioethics, most of us strive for a humane compromise, and think it's helpful to debate the pros and cons before legislation is firmed up. But there are issues where adherents of a religion adopt an absolutist stance. Even here, however, we can ensure that the facts are presented fairly, and appreciated by all those involved. For instance, in the UK parliament, during debates on embryo research and the technology of 'three-parent families', some Roman Catholics produced banners depicting embryos as homunculi with human features – whereas 14 days after gestation they're still a seemingly amorphous and undifferentiated microscopic group of cells. Scientists should surely point this out, and attempt to counter any manifest factual misperception. Of course, if a scientifically informed Catholic were to concede that point, but still argue that the embryo should be inviolate because it has a 'soul' from conception, then there can be no scope for reconciliation – nor indeed for further worthwhile debate. There is now, by the way, pressure from some biologists to extend and relax the 14-day limit on embryo research. Sadly, I doubt that this will generate such a broad consensus as Warnock achieved.

Back in the 1990s, unjustified scares about the MMR (measles, mumps and rubella) vaccine deterred parents from getting their children immunized. And sadly, we are still witnessing disturbing incidents of science denial today – there are vocal groups of anti-vaxxers in most countries. The emergence of Covid-19 in early 2020 was an emergency, where scientific advice was crucial, but had initially to be offered 'on the hoof', on the basis of limited knowledge and without time for full deliberation. The media prominence of the Covid-19 debates ever since then has given all citizens a perspective on how some uncertainties were gradually resolved by new data, whereas others, such as the actual origin of the virus, may always remain. There will be more to say about Covid-19 in section 3.1.

When we talk about science denial, we must of course acknowledge that all scientific theories are, at some level, tentative and provisional. They need to be reappraised as fresh experimental and observational data come in. But only science can promise a continuous mid-course self-correction. When I was young, milk and eggs were deemed 'good'; a decade later we were warned off them because of cholesterol; but today they seem to be OK again (at least not in excess). But we shouldn't be surprised about these shifting guidelines: there's no reason to expect scientific issues to be straightforward, even if they refer to something 'everyday' and familiar.

Science remains the best tool for discovering and under-standing facts about the cosmos and about ourselves. The key point is that it is never a good idea to bet against the judgement of science. To do so when human life (as in the case of the Covid-19 pandemic) or the future of the planet's biosphere (as in the case of climate crisis) is at stake, is unconscionable.

We live in a world where more and more of the decisions confronting governments involve scientific evidence. Obviously, pandemics and climate change have been at the forefront of our minds recently, but policies on health, energy and the environment all have a scientific dimension. However, these policies have economic, social and ethical aspects as well. And on those aspects, scientists speak only as citizens.

Yet if public debate is to rise above mere sloganeering, everyone needs to have enough of a 'feel' for science to avoid becoming bamboozled by propaganda and bad statistics. The need for proper debate will become even more acute in the future, as the pressures on the environment and those resulting from misdirected technology get more diverse and threatening. In this respect, one of the most frightening outcomes from the era of Trump and his ilk has been the death of facts. In today's 'post-truth era', when there is little agreement on what defines reliable sources, we may have to take inspiration from Galileo who (apocryphal as it may be) was reported, after being forced to deny that the Earth orbits the Sun, to have muttered 'And yet it moves', as a rallying cry: a reminder that in spite of what you may believe, the facts always remain the same.

We have been there before. In *The Origins of Totalitarianism*, philosopher Hannah Arendt prophetically wrote: 'The ideal subject of totalitarian rule is not the convinced Nazi or the convinced Communist, but people for whom the distinction between fact and fiction (i.e., the reality of experience) and the distinction between true and false (i.e., the standards of thought) no longer exist.'

2.4 Science and the media

With regard to the successful communication of science to the lay public, Darwin's *The Origin of Species*, published in 1860, is an exemplar.[3] The book was a bestseller: readily accessible – even fine literature – as well as an epochal contribution to science. But that was an exception. In glaring contrast, Gregor Mendel's 1866 paper entitled 'Experiments with Plant Hybrids', reporting the classic experiments on sweet peas conducted in his monastery garden, was published in an obscure journal and wasn't properly appreciated for decades. (Darwin had the journal in his library, but the pages remained uncut. It is a scientific tragedy that he never absorbed Mendel's work, which laid the foundations for modern genetics.)

It is unlikely that any twenty-first-century breakthroughs can be presented to general readers in such a compelling and accessible way as Darwin's ideas were. The barrier is especially high when ideas can be fully expressed only in mathematical language: few read Einstein's original papers, even though his insights have permeated our culture. Indeed, that barrier already existed for mathematical science in the seventeenth century. Newton's great work, the *Principia*, highly mathematical and written in Latin, was heavy going even for his distinguished contemporaries like Halley and Hooke; certainly a general reader would have found it impenetrable, even when an English version appeared. Popularizers later distilled Newton's ideas into more accessible form: as early as 1736 the Venetian scientist Francesco Algarotti published a book, quickly translated into English, with the title *Neutonianismo per le dame* (*Newtonianism for Ladies*).

What makes science seem forbidding is the technical vocabulary, the formulae, and so forth. Despite these impediments, the essence (albeit without the supportive arguments) can generally be conveyed by skilled communicators. It is usually necessary to eschew equations, but that by itself is not enough. Specialist jargon, which is especially impenetrable in biomedical topics, must be avoided. It is easy for scientists to forget, too, that their use of familiar words (like 'degenerate', 'strings' or 'colour') in special contexts different from their everyday usage can be baffling as well. And, to take another example, climate scientists calculate that the warming directly due to extra CO_2 can be amplified by the effects of enhanced water vapour, changing cloud cover, and so forth. This dangerous process (mentioned in section 1.2) is termed 'positive feedback' – a phrase which for many readers would have benign connotations (it's what you hope for after a job appraisal, for instance).

So the gulf between what is written for specialists and what is accessible to the average reader is widening. Literally millions of scientific papers are published, worldwide, each year. They are addressed to fellow specialists and typically have very few readers. This vast primary literature needs to be sifted and synthesized, otherwise not even the specialists can keep up. Moreover, professional scientists today are depressingly 'lay' outside their specialisms – certainly a contrast to the seventeenth-century polymaths who founded the Royal Society. My own knowledge of recent biological advances, such as it is, comes largely from excellent 'popular' books and journalism. Science writers and journalists do an important job, and a difficult one. I know from bitter experience how hard it is to explain in clear language even something I think I understand well. But journalists have the far greater challenge of assimilating and presenting topics quite new to them, often to a tight deadline; broadcasters may be required to speak at short notice, without hesitation, deviation or repetition, before a microphone or TV camera.

Science only earns newspaper headlines, or prominence in TV bulletins, as background to a natural disaster, or health scare, rather than as a story in its own right. Until Covid-19 struck, this happened rarely. Scientists shouldn't complain about this any more than novelists or composers would complain that their new works don't make the news bulletins. Indeed, coverage restricted to 'newsworthy' items – newly announced results that carry a crisp and easily summarized message, or of course spectacular feats like a probe landing on Mars – distorts and obscures the way science normally develops.

Scientific themes are better suited to documentaries and features. The terrestrial TV channels offer large potential audiences, but commercial pressures, and concern that the viewers may channel-surf before the next advertising break, militate against programmes like Jacob Bronowski's classic 'The Ascent of Man' (BBC) or Carl Sagan's 'Cosmos' (PBS) – two classic 13-part series commissioned in the 1970s. It is fortunate that big-budget commercial channels like Netflix, Discovery and Amazon have entered this arena. Even the great David Attenborough defected (at least partially) from the BBC. But there continue to be excellent programmes fronted by Jim Al Khalili, Alice Roberts, Neil Tyson and others; the internet opens up niches for more specialized content – webcasts and podcasts flourish.

I would derive less satisfaction from my astronomical research if I could discuss it only with professional colleagues. I enjoy sharing ideas, and the mystery and the wonder of the universe, with non-specialists. Moreover, even when we do it badly, attempts at this kind of communication are salutary for scientists themselves, helping us to see our work in perspective. As already emphasized, researchers don't usually shoot directly for a grand goal. Unless they are geniuses (or unless they are cranks) they focus on timely, bite-sized problems because that's the methodology that pays off. But it does carry an occupational risk: we may forget, when we focus sharply on one problem, that we're wearing blinkers and that our piecemeal efforts are only worthwhile insofar as they're steps towards addressing some fundamental question.

In 1964, Arno Penzias and Robert Wilson, radio engineers at the Bell Telephone Laboratories in the US, made, quite unexpectedly, one of the great discoveries of the twentieth century: they detected weak microwaves, seemingly pervading all of space, which are actually a relic of the Big Bang. But Wilson afterwards remarked that he was so focused on the mundane tasks of optimizing his apparatus, which even included scraping the equipment free of pigeon droppings, that he didn't himself appreciate the full import of what he'd done until he read a 'popular' description in *The New York Times*, where the paper's leading science writer, Walter Sullivan, described the background noise in their radio antenna as the 'afterglow of creation'.

Not all who make discoveries are as lucky in their 'laudator' as Wilson was. The best way to ensure that one's views get through undistorted is via one's own written words, in articles or books. Some distinguished scientists have been successful authors. But most of us dislike writing – though present-day students are far more fluent (if not more literate) than my own pre-email and pre-blog generation ever were. And unless we're very lucky, we'll reach fewer readers by writing ourselves, than if our views are amplified by a 'media star'. Many of the most successful writers of scientific books are interpreters and synthesizers rather than active researchers. Bill Bryson, for instance, has marvellously conveyed his zest and enthusiasm for 'nearly everything' and 'the human body' in two books[4] that have been read by millions. And the polymath journalist Philip Ball has written books on themes ranging from quantum weirdness to ancient Chinese waterworks, as well as an amazing range of articles.

Incidentally, scientists habitually bemoan the meagre public grasp of our subject. (And it's indeed dismaying that, in a regular poll conducted in the US, 2020 was the first year in which more than 50 per cent of Americans accepted Darwin's theory![5]) But maybe we protest too much. On the contrary, we should, perhaps, be gratified and surprised that there's wide interest in such remote topics as dinosaurs, the Large Hadron Collider in Geneva, the James Webb Space Telescope, or alien life. It is indeed sad if some citizens can't distinguish a proton from a protein; but equally so if they are ignorant of their nation's history, or are unable to find Taiwan or Ukraine on a map – and many can't.

Misperceptions about Darwin or dinosaurs are an intellectual loss, but no more than that. In the medical arena, however, 'fake news' could be a matter of life and death. Hope can be cruelly raised by claims of miracle cures;

exaggerated scares can distort health-care choices, as happened with the Covid-19 vaccines. When reporting a particular viewpoint, journalists should clarify whether it is widely supported, or whether it is contested by 99 per cent of specialists – as were the claims about the dangers of the MMR vaccine. Noisy controversy need not signify evenly balanced arguments. Of course, the establishment is some-times routed and a maverick vindicated. We all enjoy seeing this happen, but such instances are rarer than is commonly supposed. The best scientific journalists and bloggers are plugged into an extensive network that should enable them to calibrate the quality of novel claims and the reliability of sources.

Scientists should themselves expect media scrutiny. Their expertise is crucial in areas that fascinate many of us, and matter to us all. This happened on an unprecedented scale during the pandemic. As a community working under extreme pressure, they acquitted themselves well – disagreeing respectfully, and (unlike their political masters) cooperating openly across national boundaries. And they shouldn't be bashful in proclaiming the overall promise that science offers – it's an unending quest to understand nature, and essential for our survival.

There is a strong tradition of science journalism. But there is an impediment: these dedicated journalists are up against the handicap that few in top editorial positions have any real background in science. The barrier between the 'two cultures' of science and the arts that the writer and physical chemist C. P. Snow identified 60 years ago, and which was indeed prevalent in the cloistered academic society of his time, has now been at least somewhat eroded – some would argue that these 'cultures' are spanned by the social sciences. But there is still a 'divide'. The editors of even the so-called highbrow press feel they cannot assume that their readers possess the level of scientific knowledge that we might hope all school leavers would have achieved, whereas in the same media economic articles are often quite arcane, and critics of classical music (sadly a dying breed) would be thought to be insulting their readers if they explained what a concerto was. About half of the readers of the quality press are likely to have some post-school scientific education, or to be engaged in work with a technical dimension, while it is those who control the media who are overwhelmingly lacking in such basic knowledge. Of course, this is a legacy of the excessive specialization – and excessive stratification – of traditional education, on which I comment further in chapter 4.

A similar 'culture gap' is faced by scientists who serve as political advisors. Certainly the UK and US would benefit from more technically trained people in politics. But I don't go along entirely with those who would advocate a government of technocrats, as in China or even Singapore. Such people's expertise may be profound, but no individual can span more than a subfield of science. Indeed, I'd definitely prefer our university-educated senior ministers to have studied history rather than dentistry. In the UK, I've observed that the individual politicians and 'opinion-formers' who have been most effective in influencing science policy have frequently been 'generalists'.

2.5 Science's limits and our twenty-first-century challenges

The public culture of science is something that should concern us all. The era of glad optimism about science perhaps peaked soon after the Second World War. In the US the classic 1945 report 'Science: The Endless Frontier', by Vannevar Bush, presented President Truman with the case for public funding of science across a broad front. In the UK, scientists and engineers who had made crucial contributions to the Second World War, and who had, in the words of C. P. Snow, 'the future in their bones', roamed the 'corridors of power'. They influenced, among others, Harold Wilson. When he became the UK's Prime Minister in 1964, Wilson delivered a celebrated speech extolling 'the white heat of the technological revolution'.

It is interesting to speculate what would be the highlights if the UK were to host, this decade, a follow-up to the Great Exhibition of 1851, or the Festival of Britain in 1951. A paper by the historian of innovation Anton Howes speculates on this:

It would be a place where visitors would actually get to see drone deliveries in action, take rides in driverless cars, experience the latest in virtual-reality technology, play with prototype augmented-reality devices, witness organ tissue and metals and electronics being 3D-printed, and watch industrial manufacturing robots in action. They could have a taste of lab-grown meat at the food stalls, meet cloned animals brought back from extinction, perform feats of extraordinary strength wearing the same exoskeletons used in factories, fly in a jet-suit, and listen to panel interviews with people who have experienced the latest in medical advancement. Perhaps a commercial space launch using the latest technology might be timed to coincide with the event, to be livestreamed on a big screen for all visitors to see.[6]

This would be spectacular indeed, and surely energize an awed public. However, the impact of new breakthroughs is now viewed with ambivalence rather than with the enthusiasm that prevailed in the 1950s and 1960s. The advances of science since then, though the basis of marvellous technologies, create new hazards and raise new ethical issues. Indeed, many are anxious that science is 'running away' so fast that neither politicians nor the lay public can assimilate or cope with it. The stakes are indeed getting higher: science offers huge opportunities, but future generations will be vulnerable to technologies powerful enough to jeopardize the very survival of our civilization. The greatest challenges – misuses of bio- and cyber-technology – were briefly addressed in chapter 1; they concern me as a scientifically aware citizen, and as a worried member of the human race.

Arthur C. Clarke noted that any sufficiently advanced technology is indistinguishable from magic. We can't now envision what artefacts might exist centuries hence, any more than a Roman could imagine today's satnav and smartphones. Nevertheless, a physicist would unhesitatingly assert that some conjectured innovations will remain for ever fiction. For instance, from my perspective as a research scientist (an astronomer) I'd confidently say this about time machines. That's because changing the past would lead to paradoxes – infanticide would violate logic as well as ethics if the victim was your grandmother in her cradle. So, what is the demarcation between concepts that seem crazy now but might be realized eventually, and things that are for ever impossible? Are there limits to how much we can ever predict? Are there scientific problems that will forever baffle us – phenomena that simply transcend human understanding?

Einstein averred that 'the most incomprehensible thing about the universe is that it is comprehensible'. He was right to be astonished. Our minds evolved to cope with life on the African savannah, but they can also comprehend the microworld of atoms, and the vastness of the cosmos. We marvel at the fact that the universe isn't anarchic – that atoms obey the same laws in distant galaxies as in the lab. Our cosmic horizons have vastly enlarged. Our Sun is one of a hundred billion stars in our Galaxy, which is itself one of many billion galaxies in range of our telescopes. And this entire panorama emerged from a hot, dense 'beginning' nearly 14 billion years ago. Some inferences about the early universe are as evidence-based as anything a geologist might tell you about the history of our Earth; we can make confident and precise statements about how hot and dense things were in the first few seconds of our universe's expansion, even just a microsecond after the Big Bang. But, as always in science, each advance brings into focus some new questions that couldn't previously have even been posed. We now confront the mystery of the very beginning (if indeed there was one).

It is perhaps self-indulgent to focus on remote and speculative topics. But the bedrock nature of space and time, and the structure of our entire universe are surely among science's great 'open frontiers'. They exemplify intellectual domains where we're still groping for the truth – where, in the fashion of ancient cartographers, we must still inscribe 'here be dragons'. A unified theory, if achieved, would complete a programme that started with Newton, who identified the universal force of gravity, and continued through Faraday and Maxwell, who showed that electric and magnetic forces were intimately linked, and their successors. It might even realize the Pythagorean vision of reducing all nature's complexities to geometry. Until we have such a theory, we won't understand one of the deepest mysteries that astronomy has revealed – that there is 'dark energy' latent even in empty space, which pushes galaxies apart at an accelerating rate. And our successors will need to address questions that we can't yet formulate – Donald Rumsfeld's famous 'unknown unknowns' (what a pity, incidentally, that Rumsfeld didn't become a philosopher full-time!).

Einstein himself worked on an abortive unified theory till his dying day; in retrospect it is clear that his efforts were premature – too little was then known about the forces and particles that govern the subatomic world. Cynics have said that he might as well have gone fishing from 1920 onwards; but there's something rather noble about the way he persevered, reaching beyond his grasp. (Likewise, Francis Crick, a driving intellect behind molecular biology, shifted, when he reached 60, to the 'Everest' problems of consciousness and the brain even though he knew he'd never get near the summit.)

The cumulative advance of science requires new technology and new instruments – in symbiosis, of course, with theory and insight. The Large Hadron Collider (LHC) at CERN in Geneva is the world's biggest scientific instrument. Its completion in 2009 generated razzmatazz and wide public interest; but at the same time questions were understandably raised about why such a large investment was being made in seemingly recondite science. But what is special about this branch of science is that its practitioners in many different countries have chosen to commit most of their resources over a timespan of nearly 20 years to construct and operate a single vast instrument in a Europe-led collaboration. The annual contribution amounts to only about 2 per cent of the overall budget of European nations for academic science, which doesn't seem a disproportionate allocation to a field so challenging and fundamental. The only other single scientific project to match the scale of the LHC is the James Webb Space Telescope (JWST), launched in 2021. This is primarily a US (NASA) project, though with participation from Europe and from Canada. It will probe deeper into space, and therefore further back into the past, than previous telescopes, revealing how and when the first stars formed. It will also detect the faint infrared radiation from planets orbiting nearby stars, seeking spectral evidence for life on them. Successful international collaboration on immensely challenging projects like the LHC and the JWST, to probe some of nature's most fundamental mysteries – and push technology to its limits – is surely something in which our civilization can take pride.[7]

And other 'big science' ventures – vast telescopes on the ground, computer networks, etc. – are as impressive as accelerators and space probes. But for me, the most astonishing has been an instrument called LIGO: a set of lasers bouncing light between mirrors fixed 4 km apart. They're configured to detect tiny changes in the separation between the mirrors induced when a gravitational wave – a 'ripple' in space – passes across them. LIGO came online in 2015. Gravitational waves are generated by extreme astronomical events – stellar explosions or collisions between black holes. They're a crucial and distinctive prediction of Einstein's general relativity. The predicted amplitude when such transmissions reach the Earth is infinitesimal – equivalent to a stretching by the thickness of a hair in a string long enough to reach Alpha Centauri: one part in a billion trillion. The amazing US-led international project to study gravitational waves is an extreme instance when the designers and builders of the instruments deserve more credit than the theorists.[8]

We are witnessing stronger links between two frontiers of science: the very large (the cosmos) and the very small (the quantum). But only a tiny proportion of researchers are cosmologists or particle physicists. Ninety-nine per cent of scientists deploy their efforts on a third frontier: the very complex. Our everyday world presents intellectual challenges just as daunting as those of the cosmos and the quantum. It may seem incongruous that the LIGO experimenters can make confident statements about black holes colliding a billion light years away, while medical scientists are baffled about issues close at hand that we all care about – diet and common diseases, for instance. But this is because our 'living environment' is so immensely complicated. Even the smallest insect, with its layer upon layer of intricate structure, is far more complex than either an atom or a star.

It's a standard metaphor to liken the different sciences to successive levels of a tall building: physics on the ground floor, then chemistry, then cell biology, and all the way up to psychology – with the economists in the penthouse. There is a corresponding hierarchy of complexity: atoms, molecules, cells, organisms, and so forth. But the analogy fails in a crucial respect. In a building, insecure foundations imperil everything above; but the 'higher level' sciences dealing with complex systems aren't imperilled by an insecure base. The uncertainties of subatomic physics are irrelevant to biologists and environmentalists. To those who study how water flows – why it goes turbulent, or why waves break – it's irrelevant that water is molecules of H_2O. An albatross returns to its nest after wandering 10,000 miles in the southern oceans – and it does this predictably. But it would be impossible, even in principle, to calculate this behaviour 'bottom up' by regarding the albatross as an assemblage of atoms.

Everything, however complicated – breaking waves, migrating birds, and tropical forests – obeys the equations of quantum physics and is made of atoms. Most of us are 'reductionists' in the sense that we think the properties of complex systems are 'emergent' (that they arise from the interactions among the smaller building blocks of those systems, rather than requiring an extra 'vital essence'). But even if those equations could be solved for complex macroscopic entities, they wouldn't offer the enlightenment that scientists seek. Each science has its own autonomous concepts and laws that yield real insight and predictive power: reductionism isn't the route to understanding nature's complexities. Problems in biology, and in environmental and human sciences, remain unsolved because it's hard to elucidate their complexities, not because we don't understand subatomic physics well enough.

If I were to conjecture where the scientific cutting edge will advance fastest, I'd plump for the interface between biology, computing and engineering. Practitioners of synthetic biology can not only 'read out' the sequence of a genome, but aim soon to design new ones. And another burgeoning discipline, nanotechnology, aims to build up inorganic structures atom by atom, leading to even more compact devices that will enhance computer processing and memory, and could allow robots of microscopic size (which could, for instance, navigate through our blood vessels). Computers are already transformational, especially – as I noted earlier – in fields like astronomy and climate science where we can't do actual experiments. Just as video games get more elaborate as their consoles get more powerful, these 'virtual' experiments become more realistic as computers advance. And quantum computers could be a game-changer for some problems.

Some phenomena, like the orbits of the planets, can be calculated far into the future. But such cases are actually atypical. In most contexts, there's a fundamental limit to how far ahead we can predict. That's because tiny contingencies have consequences that grow exponentially: the classic example given by 'chaos theorists' is that a storm occurring in the northern hemisphere can depend on whether a butterfly flapped its wings in South America. For reasons like this, even the most fine-grained computation cannot normally forecast British weather more than a few days ahead. (But – importantly – this doesn't stymie predictions of long-term climate change, nor weaken our confidence that it will be colder next January than it is in July.) So there are limits to what can ever be predicted about the fine detail of the future, however powerful computers become.

I've offered some predictions in chapter 1, but is there more that we can conjecture about developments in the rest of this century and beyond? Number one on my list would be the future of humanity itself. One thing that's changed little for millennia is human nature and human character.

Before long, however, new cognition-enhancing drugs, genetics, and 'cyborg' techniques may alter or 'enhance' human beings themselves. That's something qualitatively new in recorded history – and disquieting because it could portend more fundamental forms of inequality if options such as life extension (see section 1.3) were open to only a privileged few. One hopes that such changes will be effectively regulated and constrained. However, this prospect is relevant to another futuristic scenario: that some humans may spread beyond the Earth and attempt to settle on (for instance) Mars. They would be ill adapted to the hostile Martian environment – and beyond the clutches of the regulators – so it's these adventurers who may have the motives and the freedom to exploit these advancing biotechnologies. The longer-term future is the province of science fiction writers. And I advise students to read first-rate SF rather than second-rate science (natural or social). It's more entertaining, and no more likely to be wrong.

2.6 Science is a team game

In any good SF, the main protagonists matter. And so what of scientists themselves, the 'scientific tribe'?

When children (or cartoonists) want to depict an archetypal scientist, they frequently draw a wild-looking, male figure, resembling the familiar image of Einstein. I recall a cartoon showing Einstein standing in front of a blackboard. On it is written $E = ma^2$ – crossed out; then $E = mb^2$ – also crossed out; and then, $E = mc^2$ – eureka! Of course, that's not how it happened. Einstein's annus mirabilis was 1905. As well as discovering the equation that everyone knows, he wrote three other papers that year, any one of which would have established his reputation. He was then 26 years old, working from nine to five in the Swiss patent office as a 'technical expert, 3rd class'. Pictures from that time show him as a rather dapper young man. But it's the old Einstein who's become the iconic figure – the benign and unkempt sage of posters, T-shirts and cartoons.

In impact on our perception of the physical world Einstein is matched only by Isaac Newton (and, in the biological sciences, of course, by Darwin). But in charisma, there's no contest. Newton was an unappealing character: solitary and reclusive when young; vain and vindictive in his later years. Einstein would have been better company. It's fortunate that the most famous twentieth-century scientist projected an engaging image: a genial figure, ready with an aphorism, and idealistically engaged with the world's problems.

But the popular perception of Einstein has a downside. It unduly exalts 'armchair theory'. Science wouldn't have got far by pure thought alone: we're no wiser than Aristotle was. It has developed in symbiosis with advancing technology, from telescopes to computers. 'Lone thinkers' are indeed one category of scientist. But they're a small minority: there is a variety of working styles, which mostly involve more of a 'team effort'.

The greatest scientists, even the solitary ones, don't fall into a single mould. Some are brilliant. Newton's mental powers seem to have been really off the scale. His concentration was as exceptional as his intellect: when asked how he cracked such deep problems, he said 'by thinking on them continually'. In contrast, Darwin was modest in his self-assessment: he wrote: 'I have a fair share of invention, and of common sense or judgement, such as every fairly successful lawyer or doctor must have, but not, I believe, in any higher degree.'

Science involves hard slog, but that's not enough: insights – 'eureka moments' – are crucial, too. There are parallels with creativity in the arts; but there are differences, too. Any artist's work is individual and distinctive but it generally doesn't last; contrarywise, even the journeyman scientist adds a few durable bricks to the corpus of public knowledge. But our scientific contributions lose their identity: if A didn't discover something, in general B soon would – indeed, there are many instances of near-simultaneous discovery. Not so, of course, in the creative arts. On this topic, I like to quote the great biologist Peter Medawar's remark that when Wagner diverted his energies for ten years, in the middle of the Ring cycle, to compose Die Meistersinger and Tristan, he wasn't worried that someone would scoop him on Götterdämmerung.

Even Einstein exemplifies this contrast. He made a greater and more distinctive imprint on twentieth-century science than any other individual; but had he never existed, all his insights would by now have been revealed, though gradually, and probably by several people rather than by a single great mind. Einstein's fame extends far beyond science; he was one of the few in the field who really did achieve public celebrity and is as much an icon of creative genius as Beethoven. His impact on general culture, however, has been ambivalent. It's a pity, in retrospect, that he called his theory 'relativity'. Its essence is that the local laws are just the same in different frames of reference. 'Theory of invariance' might have been a more apt choice, and would have staunched the misleading analogies with relativism in human contexts. But in terms of cultural fall-out, he's fared no worse than others. Heisenberg's uncertainty principle – a mathematically precise concept, the keystone of quantum mechanics – has been hijacked by adherents of oriental mysticism. And Darwin has likewise suffered tendentious distortions, especially in applications to human psychology.

Engineering of course is not a 'solitary' pursuit though it plainly has its iconic figures. It's good that so many have heard of great nineteenth-century engineers like Telford, Brunel and Edison. They might be harder pressed to think of engineers who flourished in the twentieth – maybe those in the UK would come up with Frank Whittle, Tim Berners-Lee or James Dyson. (Indeed, the engineering profession has been even worse at PR than academic scientists are, otherwise their leading practitioners should surely have the same glamorous profile as our most celebrated architects.) In particular, those who've given us today's amazing technologies deserve greater acclaim. Fortunately, there is a transformation in the present century. The ascendancy of Microsoft and Apple, of Google and Facebook, bears witness to how a cohort of young scientific entrepreneurs have changed the world. And through his large-scale innovations, seriously disrupting the vast conglomerates of the motor and aerospace industries, some argue that Elon Musk deserves to rank as a twenty-first-century Brunel.

This focus on iconic figures from the past misleadingly typecasts academic scientists as eccentric, elderly white men – some of us are, but, luckily, we're a diminishing fraction of a far more varied community, though there is a long way to go before an acceptable range of diversity is achieved. Apart from issues of fairness and justice, it's obviously an 'own goal' to exclude any group of talented individuals from the scientific enterprise.

In astronomy – the science I know best – there was nothing like a 'level playing field' for women until the end of the twentieth century. There are famous historical cases of women being under-recognized. William Herschel, who mapped the Milky Way and discovered Uranus in the eighteenth century, had a lot of help from his sister Caroline. A century later William Huggins, who discovered spectroscopically that stars contained the same chemical elements as the Earth (not some mysterious 'fifth essence') and became president of the Royal Society in his old age, was assisted for decades by his wife, Margaret. In the 1920s Cecilia Payne wrote one of the most important ever PhD theses in astronomy, showing that the Sun and stars were mainly made of hydrogen. Her work was received with undue scepticism, but she eventually became the first female full professor at Harvard (in any subject). And, nearer our own day I knew Geoffrey and Margaret Burbidge – two eminent astronomers. He was a theorist and she was an observer. But in the 1950s women weren't allowed on the big Californian telescopes so the time had to be allocated to Geoff, with Margaret cited as his 'assistant'. But she went on to a much-admired career and in 2019 died aged 100. I'm glad to say that among the younger generation the proportion of women is higher in astronomy than in the rest of the physical sciences. But it's still low in fields like computer science – and equally so in the top ranks of more traditional businesses.

Science is, for most of us, an intensely interactive activity. Certainly, what I've enjoyed most in my own career has been involvement in ongoing debates that have gradually clarified perplexities, and expanded the area of consensual understanding. It is important, as well as enlightening, to appreciate how pervasive the social and political factors that drive and direct science are. The way scientists work, the institutions that support them, what problems attract their interest, what styles of explanation are culturally appealing – and (more mundanely) what fields attract funding – plainly depend on a range of political, sociological and economic factors. These change over time, and create different pressures and demands in different countries. Some projects, especially big international ones like the space programme, are a by-product of activities driven by other imperatives. However – and this is crucial – the outcome of scientists' efforts is objective: it can be evaluated by criteria that don't depend on how these ideas were motivated and arrived at. How science is applied, however, is a culture-dependent matter.

Steven Weinberg was one of the greatest theoretical physicists of his generation. But he was, more than that, unusual in being also an intellectual and a fine writer. His book *Dreams of a Final Theory* gave an apt metaphor for scientific breakthroughs:

> A party of mountain climbers may argue over the best path to the peak, and these arguments may be conditioned by the history and social structure of the expedition, but in the end either they find a good path to the summit or they do not, and when they get there they know it.[9]

By analogy, it is fascinating to study how social and economic factors moulded the development of music – for instance, the shift from liturgical to operatic genres; the increase in the scale of orchestral compositions after the transition from private patronage to public concerts, and so on. But such studies, though worthwhile in their own right, are peripheral to the essence of the music itself.

I end this chapter with some personal comments. When I took stock on reaching the age of 60, I didn't feel I could claim a really major scientific contribution, but felt fortunate to have been an active participant in scientific debates, and to have closely followed (and sometimes influenced) the collective progress of projects and concepts that will surely be recognized, when the history of science is written, as providing one of its most exciting chapters. And to have done this while based in a near-ideal academic environment, with colleagues and collaborators (both in Cambridge and spread around the world) who, with hardly any exceptions, I respected and enjoyed interacting with. I've been part of an unusually fortunate generation. Younger colleagues are pressured to have a narrower scientific focus, to have a more competitive attitude, and to be enmeshed in a more vexatious bureaucracy.

But I'd noticed three ways in which scientists grow old. Some lose interest in research, turning to other activities (or just lapsing into torpor). Others continue their research: they realize that (in contrast to many composers and artists) their last works are unlikely to be their best, but are content to continue 'on a plateau', doing what they're good at. What we get worst at as we age is absorbing new influences and mastering new techniques. Science is a social and collective enterprise and its practitioners have to do this to remain 'cutting edge'; there's a contrast with artists, who can deepen their creativity through internal development alone. But there is a third category, which includes many of science's major figures (including, disconcertingly, Fred Hoyle and Arthur Eddington, two previous holders of my professorship!). These people retain their research motivation: they would claim they are still trying to understand the world. But they no longer get satisfaction from continuing routine work in their own field. They instead make incursions into other fields of knowledge where they lack expertise; through arrogance they over-reach themselves and embarrass their admirers, if not themselves. They lose their critical faculties even if they retain others.

Mindful of these pitfalls, I felt it would be prudent to 'diversify', in the hope that I could, maybe for another decade, cope with new challenges. I therefore had several reasons for wanting to take on something additional to research (and something that might be more of a service than popular writing, at which I was far less adept than many others). Here I was lucky – in a sense I was too lucky, in that by seizing several opportunities I 'overshot', taking on so much extraneous work that I ended up doing even less research during the decade that followed than I'd have wished if I could have apportioned my time optimally. The later parts of this book are somewhat more 'parochial' because, having been latterly involved in varied administrative and policy issues, I can illustrate my themes by examples drawn from my experience.

It is important to reiterate that science requires a variety of personality types and working styles. There are indeed 'lone thinkers'. But far more are involved in experiments, computations, data analysis, fieldwork or quasi-industrial work in big projects. Nonetheless, almost all scientists, even those whose research is solitary, are part of communities

– research institutes, universities, academic societies or industrial labs – which offer them support and help promote their work. These institutions are the subject of the next chapter.

Chapter 3

Science Comes out of the Lab

3.1 Lessons from Covid-19

Governments – politicians and their officials – need access to the best expert advice. In times of stability, this can come from national academies, from specialized professional societies, and even from the reports of government committees (for instance Congressional Committees in the US and Parliamentary Select Committees in the UK). Unless a crisis is imminent, it's hard to focus ministerial attention onto even the most important long-term policy issues (securing the electricity supply, the internet, supply chains for essential goods, resilience against extreme weather, etc.). But sometimes a catastrophic event has an urgent impact on our lives and governments need to make hard and quick decisions in the face of uncertainty, as happened on a mega-scale in 2020 when Covid-19 ravaged the world.

The emergence of Covid-19 signalled a potential global catastrophe. The Trump administration's initial response was to dismiss scientific warnings entirely. On 22 January 2020, the President said in an interview: 'We have it totally under control. It's one person coming in from China, and we have it under control. It's going to be just fine.' Then, a month later, he triumphantly declared: 'when you have 15 people [infected by the coronavirus], and the 15 within a couple of days is going to be down to close to zero, that's a pretty good job we've done.'

Even as late as 12 March, in a meeting with the Irish Prime Minister, Trump continued to brag: 'Because of what I did and what the administration did with China, we have 32 deaths at this point. Other countries that are smaller have many, many deaths.' A year later, the US had suffered well over 500,000 deaths.

In the UK, Prime Minister Johnson was slow to recognize the grave risks involved, and to implement an effective response. The advice he received on quarantine, etc., was initially criticized for being dominated by a closed group of favoured advisors – the Scientific Advisory Group for Emergencies (SAGE). Initially, SAGE's membership wasn't published. But the process quickly opened up, enabling dozens of independent voices to participate in the discussion, and sometimes to offer vehement criticism of government policy, which of course often shifted as the evidence changed and knowledge firmed up. David King, a former Chief Scientific Advisor to the UK government, even set up a rival 'independent SAGE' committee.

Even when the scientific facts are agreed upon, the planned response depends on balancing the ethics, economics and politics – and consensus isn't easy to reach even among experts. For instance, shutting schools down may reduce the spread of infection; but might not this benefit be outweighed by the harm done by disrupting children's education – especially that of disadvantaged children whose parents couldn't offer effective home schooling?

To interpret scientific advice and calibrate policy options for politicians, it's crucial to have some experts embedded in the government at a high level. In some countries Covid-19 brought these scientific advisors out of the shadows: a few became public figures – receiving some acclaim, but enduring the odium that's the fate of any prominent person in the age of social media. In the US, Anthony Fauci gained wide public trust, despite the scorn of President Trump and some conspiracy theorists. And in the UK, there were regular press conferences when the Prime Minister (or a senior substitute) was flanked by the government's Chief Scientific Advisor, Patrick Vallance, and the Chief Medical Officer, Chris Whitty. There were even some 'celebrity statisticians' prominent in the media, who helped the public to avoid being bamboozled by numbers.

We could have learnt more from some sudden unexpected crises – though not on the Covid-19 scale – that had flared up in recent decades. For instance, in April 2010, dust from the eruption of the Eyjafjallajökull volcano in Iceland disrupted air travel throughout northern Europe and raised urgent questions about vulcanology, about wind patterns, and about how different kinds of dust affect jet engines. In that instance, the knowledge was basically there; what was lacking was coordination and an appropriate protocol (clarifying, for instance, the terms of the warranty on the planes' engines). The result was an overcautious reaction: shutting down all air travel in northern Europe. And, of course, there are more frequent localized emergencies – fires, floods, electric grid failures, foot and mouth disease outbreaks, and the like – where officials need expert advice, and where the consequences will be less severe if such scenarios have been studied and planned for in advance.

Sometimes, though, even the key basic science isn't known – not even to the extent that key features of the Covid-19 threat were. An example was the outbreak of BSE or 'mad cow disease' in the UK in the 1980s. At first, experts conjectured that this disease posed no threat to humans because it resembled scrapie, which had been endemic in sheep for 200 years without crossing the species barrier.[1] That was a reasonable conjecture, and comforting to politicians and public; but it proved wrong. The pendulum then swung the other way. The scientists were braced for more than 100 deaths. But if asked 'Is the chance of a million deaths below 1 per cent?', they couldn't have confidently answered 'Yes'. 'Beef on the bone' was banned, for instance: this was, in retrospect, an overreaction, but at the time seemed a prudent precaution against a potential tragedy that could have been far more widespread than it turned out to be. (The actual impact of this disease, though attenuated, has sputtered on for decades: in 2021, there were two fatalities among French laboratory workers.)

Governments, including the UK's, had the prudence to stock up vaccines against swine flu in 2009 – even though it turned out, fortunately, that recent influenza epidemics have been milder than feared. But the threat of a coronavirus pandemic – where protective clothing for health workers was needed, and there was no guarantee that there would be a vaccine – was not adequately prepared for. If we apply to pandemics the same prudent analysis whereby we calculate an insurance premium – multiplying probability by consequences – we'd surely conclude that measures to alleviate this kind of extreme event should be hugely scaled up. (And these measures need international cooperation. Whether or not an epidemic gets a global grip may hinge, for instance, on how quickly a farmer in East Asia or Africa can report any strange sickness in their animals.)

However, there's a mismatch between public perception of very different risks and their actual seriousness: some induce a special 'dread factor' because we can't control our exposure to them. We fret unduly about carcinogens in food and low-level radiation. We worry more about deaths via terrorism than about the far greater number caused by road accidents. We are, in contrast, in denial – until they occur – about disruptions to society by 'high-consequence/low-probability' events like pandemics. The 2008 financial crash was another such event. And cyber-threats, failures of the electric grid, and solar storms are further contingencies for which we are under-prepared.

The wide range of topics I've mentioned show how pervasive science is, in our lives and in public policy. President Obama certainly recognized this when he filled some key posts in his administration with a 'dream team' of top-rate scientists. He opined that their advice should be heeded 'even when it is inconvenient – indeed especially when it is inconvenient'. And Joe Biden showed equal commitment when he named a top geneticist, Eric Lander, as science advisor, and a member of his cabinet. Lander's remarkable career started with a PhD in pure mathematics at Oxford, followed by a period at MIT business school, before taking up genetics and playing a key role in the human genome project and its follow-up. It was sad for US science that, early in 2022, bullying allegations led him to resign.[2]

It is indeed the elected politicians who should make decisions; scientists should be 'on tap, but not on top' (pedants might be interested that this dictum, often attributed to Churchill, can be traced back to the Irish politician George Russell, who made the remark in 1912!). But the role of scientific advice is not just to provide facts, still less to support policies already decided. Experts should be prepared to challenge decision-makers, and help them to navigate the uncertainties. Indeed, Lander's enforced departure has stimulated discussion about the appropriate advisory structure. Should the top advisor be in the Cabinet – responsible for execution of policy as well as advice, but changing with each administration? Or should advisors be sufficiently at arm's-length that they can stay in office when governments change – as in the UK?

In retrospect – and understandably – the scientific advice in the early weeks of the Covid-19 pandemic wasn't always clear: for instance, the value of mask-wearing was seriously doubted by some experts in both the US and UK. But opinion gradually converged. It took time, too, to categorize the medical consequences of the virus on infected individuals of different ages and different health-status; moreover, their prognosis (including 'Long Covid') took time to be pinned down; and the timing or location of the emergence of new variants – a driver of case numbers in every country – was plainly unpredictable, as was the speed with which vaccines could be developed, and their efficacy.

We saw during the Covid-19 pandemic how politicians attempted to 'tension' various criteria in deciding the best length and stringency of lockdowns, failing to achieve anything approaching unanimity. In the UK, the delays in following advice to impose lockdown in 2020, and to restrict inward international travel (particularly from India in April 2021, whereby the 'D variant' entered the country) were motivated by political and economic pressures and seem clear errors in retrospect. Such judgements have been even harder in the case of the hyper-transmissible Omicron variant because they had to be made before the effects of the virus on individuals had been established. There was also concern (seemingly unwarranted) about whether the public's fatigue would reduce the level of acquiescence in obeying restrictions.

And politicians confront dilemmas that involve gauging the public's ethical judgements and sensitivities. If vaccines (or hospital beds) are limited, how should trade-offs be made between 'essential' workers, the young, and the old? And should countries where 70 per cent of people are already vaccinated send the remaining vaccines to African countries where the proportion is little more than 1 per cent, or continue to prioritize their own citizens for 'booster' jabs?

Scientific advisors mustn't forget that, in domains beyond their special expertise (where their input is indeed often crucial), they speak just as citizens, with no enhanced authority. That's important not only in the context of pandemics. In policy judgements about nuclear weapons, energy, environment, drug classification or health risks, political decisions are seldom purely scientific: they involve ethics, economics, and social policies as well.

Coping with these 'tensions' is especially urgent in formulating a response to climate change (discussed in chapter 1). Even if there were minimal uncertainty about how the world's weather might change, there would still be divergent views on what governments should do about it. There's a balance to be struck between mitigating global warming and adapting to it. And there are other questions. How much should we sacrifice now to ensure that the world is no worse when our grandchildren grow old? How much subsidy should be transferred from the rich world, whose fossil fuel emissions have mostly caused the problem, to the developing nations? How much should we incentivize clean energy? Should we gamble that our successors may devise a technical fix that will render nugatory any actions we take now? On all these choices there's as yet minimal consensus, still less effective action. Climate scientists, including those advising the government, themselves have a range of opinions on what the best policies should be: but they should express these views as citizens, and not claim special weight for them.

In the past, many people unquestioningly accepted 'authorities' on most topics, but that has now changed. We can all access far more information than ever before and we want to weigh up evidence for ourselves (though we may find it hard to assess what evidence is authoritative and what is 'fake' or biased). Such scrutiny should be welcome; just as there are instances of shoddy work, error or even malpractice in the medical and legal professions, so there are in science.

The most intractable crises – whether medical or climatic – have an international dimension. This leads to extra complications because current practice in archiving and managing data is not uniform across all fields, nor across all countries. Nor is there a consensus on the appropriate guidelines for making such information available. We surely need to facilitate open international debate, to ensure that scientific claims are robust and firmly grounded

– even though this will render the role of the government's scientific advisors even more challenging.

Lessons can certainly be learned from Covid-19, but the relationship between science and government is complex and eludes a simple formula for success – though, if science is to save us from the threats explored in chapter 1, it is a relationship that needs to be better understood and practised. One signature area that demonstrates this particularly well is the close and longstanding relationship between science and defence.

3.2 The world of defence

National defence is a driver of high-tech advances in space technology, laser weaponry, robotic killers and so forth. But it's all ethically ambivalent – and so it's been throughout the ages. The great chemist Haber developed poison gases in World War I; Rumford's discoveries on heat were a by-product of efforts to improve artillery. Indeed, this tension can even be traced back to Archimedes, who did 'pure' geometry but also reputedly speculated about ways of focusing sunlight on enemy ships to destroy them.

Scientific expertise was diversely and massively deployed in the Second World War – most monumentally in the Manhattan Project that led to the first atomic bomb, but also in radar, operational research and code-breaking. When war ended, most scientists returned with relief to peacetime academic pursuits. But for some, especially those who had helped build the bomb, the ivory tower was no sanctuary: they continued to do what they could, however little, to control the powers they'd helped unleash. Among the most committed and idealistic of these 'atomic scientists' was Joseph (Jo) Rotblat, the instigator and driving force behind the Pugwash Conferences. I was privileged to know him in his later years.

Rotblat's life spanned the twentieth century (he lived to be 96) and was moulded by its crises and horrors. Born in Poland in 1908, he and his family suffered severe privations in the First World War. He was not only highly intelligent but also – a lifelong trait – energetic and persistent. After apprenticeship as an electrician, he managed to enrol in university; by the age of 30 he had achieved international standing in radioactivity research.

In 1939 he accepted a short-term post with the famous nuclear physicist James Chadwick in Liverpool – a fortunate move as Warsaw's Radiological Laboratory was destroyed a year later. He became Chadwick's trusted lieutenant. The feasibility of a fission weapon was being studied under the code-names 'Maud' and 'Tube alloys'. When this UK effort was subsumed into the Manhattan Project, Chadwick used his influence to enable Rotblat to move to Los Alamos, despite his Polish citizenship. Ever since 1940, nuclear physicists had feared a 'nightmare scenario' in which Hitler might develop atomic weapons. For Rotblat this was the only moral justification for the Allies' bomb project. He was, famously, the only scientist to leave Los Alamos – where he had the status of division head – when this threat seemed no longer realistic. His motives were complex, but he recounted that the trigger was a comment by General Groves, overall head of the project, that the bomb could be used against the Russians.

Rotblat returned to Liverpool, helping to rebuild Chadwick's laboratory, but moved in 1950 to a chair in radiation medicine attached to St Bartholomew's Hospital in London. He first gained public prominence in 1954. By analysing radioactive dust deposited on a Japanese fishing boat that had strayed dangerously close to an American thermonuclear test, he inferred key features of the bomb design; he featured with Bertrand Russell in a BBC programme on the hazards of nuclear fall-out. Russell subsequently prepared a manifesto. Rotblat got Einstein to sign it too; it was Einstein's last public act – he died a week later. This 'Russell–Einstein manifesto' was then signed by nine other eminent scientists. The signatories claimed to be speaking 'not as members of this or that nation, continent or creed, but as human beings, members of the species Man, whose continued existence is in doubt'.

The manifesto led to the initiation of the Pugwash Conferences – so called after the village in Nova Scotia where the inaugural conference was held, sponsored by a Canadian millionaire called Cyrus Eaton. These meetings, which continue to this day, helped to sustain a dialogue between scientists in Russia and the West throughout the Cold War. When the Pugwash Conferences were recognized by the 1995 Nobel Peace Prize, half the award went to the Pugwash organization, and half to Rotblat personally, as their prime mover and untiring inspiration. (And it was characteristic of his austerity and dedication that he donated his half to the organization too.)

How much actual influence did these non-governmental gatherings have? During the 1960s, Pugwash Conferences undoubtedly offered crucial 'back channel' contact between the US and the Soviet Union, thereby easing the path to the partial test ban treaty of 1963, and the subsequent Anti-Ballistic Missile and non-proliferation treaties. In later decades, the Pugwash agenda broadened to include biological weapons and problems of the developing world; and other channels opened up. The conferences thereby became less distinctive and their influence was diluted. Rotblat himself retained his focus on the long-term aim to rid the world completely of nuclear weapons. This view was widely derided as woolly idealism. But it gradually gained broader 'establishment' support. For instance, in his later years, Robert McNamara, US Defense Secretary during the Cuba crisis, attended several Pugwash meetings. This might have seemed incongruous, just as Rotblat's friend-ship with Mikhail Gorbachev did. But these men converged in a realization that eliminating nuclear weapons should be an eventual goal. This view later entered the mainstream – espoused by the US 'gang of four' (Kissinger, Nunn, Perry and Schutz). In his memoirs, McNamara acknowledged that the Americans were 'lucky, as well as wise' to have avoided triggering nuclear devastation during the Cuba crisis. President Obama reactivated the disarmament agenda by persuading the US senate to ratify the New START (Strategic Arms Reduction Treaty) agreement, and in, for instance, his inspirational speech in Prague in April 2009, he enunciated 'America's commitment to seek the peace and security of a world without nuclear weapons' and pledged an array of specific steps to move in that direction. Despite hiatus during the Trump era, there is still hope.

In a memorable speech at the Royal Society, Rotblat warned scientists against exchanging the ivory tower for defence work. He quoted Solly Zuckerman, another 'estab-lishment' figure who became more 'doveish' in old age: 'When it comes to nuclear weapons . . . it is the technician who is at the heart of the arms race.'

Despite his 'nuclear' focus, Rotblat was also exercised about hazards that could stem from the misuse of twenty-first-century science. He favoured a 'Hippocratic Oath' whereby scientists would pledge to use their talents for human benefit. Whether or not such an oath would have substance, there can be no doubt of his consciousness-raising influence. He realized the need to convey his concerns beyond the Pugwash community, and to a younger generation. Even in his nineties, he could still captivate student audiences. Rotblat's inspiring life – pursued, against a backdrop of tragedy and hardship, with idealism but without illusions – deserves to be better known.[3]

None of the scientific generation with senior involvement in the Second World War remains alive today. In the US, they have been followed by an impressive cohort of scientists – people from succeeding generations who have done a spell in government, or in high-tech industry, and who serve regularly as consultants to the Pentagon or on advisory committees. In the UK, there are depressingly few younger scientists who can match the credentials and expertise of their US counterparts in providing independent expertise on nuclear issues (though there's less asymmetry in bio- and cyber-security). The reasons for this transatlantic imbalance aren't hard to find. In the US, many senior staff shuttle between government jobs and posts in, for instance, Harvard's Kennedy School or Stanford's Hoover Institute whenever the administration changes. There are always some who are 'out' rather than 'in'. The UK, in contrast, doesn't have a revolving-door system; government service is still generally a lifetime career (though this is becoming less so). For this reason, and because secrecy is more pervasive, engagement with defence issues tends, in the UK, to be restricted to a closed official world.

There is in the US, incidentally, one distinctive format for such engagement that seems to have had sustained traction. This is the JASON group. It was founded in the 1960s with support from the Pentagon to advise the US government on matters of science and technology, mostly of a sensitive nature. It involves top rank academic scientists; in the early days they were mainly physicists, but the group now embraces biologists, computer scientists and others. They're bankrolled by the Defense Department, but it's a matter of principle that they choose their own new members. Some – Dick Garwin and Freeman Dyson, for instance – remained members for more than 50 years. The JASONs spend about six weeks together in the summer, with other meetings during the year. It's a serious commitment. Its durable success has stemmed from its ability to gather a group of outstanding scientists who enjoy cross-disciplinary discourse and tossing ideas around; and who respect each other.[4]

The social chemistry of such a group hasn't been fully replicated elsewhere. It won't take off unless members dedicate substantial time to it – and unless the group addresses the kind of problems that play to their strengths, and unless members feel that their input to government decision-making is genuinely influential. I think an attempt to replicate this format in the UK would require a focus that wasn't military but addressed civilian themes such as transport, food, energy, IT and the environment, where complex systems are involved and technical innovation is needed to ensure national (indeed global) security and to confront the challenges in chapter 1. Many of us engage with government policy via committee work – sitting round a table for one-day meetings, expressing views that are then minuted by civil servants. This can sometimes be worthwhile, at least as information exchange, but it's far less impactful than a format that can generate genuinely new ideas, which is surely where the joy and excitement of science lies.

The fact that there are few well-informed independent voices weakens the quality of UK debate on defence matters (and on nuclear issues almost quenches it completely). There is likewise a transatlantic contrast with regard to defence laboratories – which are a far more closed world in the UK than in the US. I have come to know a number of people at two of the leading US labs, Livermore and Los Alamos, because their staff are encouraged to spend part of their time on research that can be presented at open meetings and published in the regular literature. 'Extreme' cosmic phenomena (supernova explosions, for instance) involve similar processes to those in bombs. Being myself an astrophysicist, I therefore follow the public and unclassified research of people from US defence labs; I can conclude with some confidence that they are competent – and that their bombs would work. But, though based in the UK, I don't really know anyone from Aldermaston, the UK's counterpart of Livermore; there are very few 'outsiders' who can vouch for their competence in the way that many of us can for their US counterparts. I wouldn't recommend any young UK physicist to join that closed world unless a more open policy were adopted, so they weren't cut off from the rest of their research community.

I have myself had very limited contact with defence issues other than as a citizen taking part in demonstrations and campaigns (and, more recently, 'backbench' contributions to parliamentary debates). One modest involvement was participation in an independent commission sponsored by the British American Security Information Council think-tank (BASIC) on 'The Future of Trident': this is the UK's so-called 'independent deterrent', consisting of four nuclear-powered submarines carrying US-made Trident missiles armed with British-made H-bombs. Our commission was co-chaired by Malcolm Rifkind, Menzies Campbell and Des Browne – three grandees of Scottish politics for all of whom I developed great respect. And the other members were more expert than me. The report didn't recommend abandoning our nuclear deterrent (as my Pugwash colleagues would have advocated). But it did open up discussion about how to further scale down the current system based on four submarines – but sadly without effect, as the UK government subsequently decided to increase the number of warheads on each submarine. We also raised awareness of new threats: for instance, whether cyber attacks could hack into the command-and-control systems for nuclear arsenals.

It was notable that I was the only one on the Trident panel with zero security clearance. Unlike some academics, I have intentionally avoided involvement in classified work. This is because it would frustratingly constrain me in informal discussions. Secrecy is more pervasive in the UK than in the US. I know a number of people in the US (from Dick Garwin[5] downwards) who have been heavily involved in the Pentagon, the JASON group, and so forth. I feel relaxed about sharing with my colleagues any 'unclassified' thoughts they express to me in informal conversations. However, it's likely that some of this information would be deemed 'secret' in the UK; I'd feel less relaxed in discussion – and more muzzled – if there could be suspicions that I was illicitly revealing facts that had been imparted to me in confidence in the UK.

Although one must accept the need for secrecy and confidentiality in some contexts, I think the amount is excessive. Within all government departments, wider scrutiny and openness would enhance the quality of 'official' advice. This is especially true for contentious issues that require specialized scientific expertise: a politician should surely prefer to receive advice which is the outcome of robust debate among all available experts.

3.3 Advisors and campaigners

Defence and arms control are a diminishing part of the agenda for today's professional scientists: as I outlined in chapter 1, the agenda is now far wider and more complex – and the issues span all of the sciences. Discussions of these are more open (though still not open enough) and often global. There is less demarcation between experts and laypersons; campaigners and bloggers enrich the debate. But professionals have special obligations to engage and men like Rotblat were inspiring exemplars.

Indeed, scientists in all fields should be prepared to divert some of their efforts towards public policy – and engage with individuals from government, business and NGOs. The great mathematician Michael Atiyah (one-time President of the Pugwash Conferences as well as of the Royal Society) offered a nice analogy: if you've got teenage children, you're a poor parent if you don't care about them and how they'll fare in adulthood, even though you've got limited influence over them. Likewise, scientists, whatever their expertise, shouldn't be indifferent to the fruits of their ideas – their creations. Their influence may be limited, but they should try to foster benign spin-offs, commercial or otherwise. They should resist, so far as they can, dubious or threatening applications of their work, and alert the public and politicians to perceived dangers. Above all, their input is crucial in formulating appropriate policies to deal with the three grand challenges highlighted in chapter 1.

Scientists in government employment, and to some extent those working for large companies, are constrained from overt campaigning. This is easier for independent entrepreneurs – and for those in academia. Indeed, universities surely have an obligation to ensure that their expertise is channelled towards issues where natural or social scientific expertise is relevant. This has of course happened during the Covid-19 response: academic scientists in many countries have been crucial in analysing virus variants, developing vaccines, modelling the spread of infections, the effects of masks and indoor ventilation, and so forth. There will, when the pressure from Covid-19 eases off, be a need to focus on other less immediate threats to ensure better preparedness.

As a modest contribution to this broad policy arena, some of us at Cambridge University founded in 2012 a Centre for the Study of Existential Risks (CSER).[6] Its aim is to focus on extreme events, especially those stemming from advancing technologies such as synthetic biology and AI. Its rationale is that, though we fret unduly about small risks, which are heavily studied, we are in denial about high consequence/ low probability events – potentially so catastrophic that even one occurrence could be one too many. These latter scenarios deserve more attention. The initiative for CSER came from Huw Price, then recently arrived in Cambridge as Bertrand Russell Professor of Philosophy, stimulated by a chance encounter with Jaan Tallinn, a co-founder of Skype, who offered seed funds.

My own interest in these issues dates back at least to my book *Our Final Century*, published in 2003 – indeed to involvement in Pugwash and inter-academy meetings in the 1980s, and demos well before that. I am keen that CSER should become an established group with wide international links. Given its strength in physical sciences, genetics and biodiversity, a university like Cambridge should use its expertise and convening power to address policy questions – in particular, to decide which global threats are real and which can be dismissed as science fiction, and to investigate how to reduce the former kind.

In the UK, a parallel activity to CSER is the Future of Humanity Institute (FHI) in Oxford. This is a spin-off from the Oxford Martin School,[7] established through a major benefaction by (the late) James Martin – whom I knew, and who was apparently influenced by my book in writing one of his own, and in endowing his Institute. The School's founder-director Ian Goldin, an economist who had held senior posts in the World Bank, had a network of international contacts (a 'Davos Man'); and his books and articles had gained him the respect of academics. He had just the right combination of qualities for the post. His successor, Charles Godfray, is a distinguished ecologist who has been active in addressing (via the UK government and the UN) issues of environment, food and conservation. There are a few university-based groups and think tanks in the US with a similar focus on 'long tail' threats. But the overall effort worldwide on these potential global disasters is still disproportionately small: the stakes are so high that, if we can reduce the probability of tech-induced catastrophe even by one part in 1,000, we'll have more than earned our keep.

Science advisors to government have the advantage of being 'insiders' but they have limited influence except in emergencies like Covid-19. They must enhance their leverage, by involvement with NGOs, via blogging and journalism, and by enlisting charismatic individuals who can use the media to amplify their voice – and of course this is still more necessary for 'outsiders'.

This is especially crucial in climate policy where I'd highlight a disparate quartet who have collectively shifted opinion and raised public awareness. The Pope's 2015 encyclical *Laudato si'* helped smooth the path towards a consensus at the Paris climate conference COP20.[8] Our secular pope David Attenborough, through his TV programmes and his speeches, has raised ocean pollution up the agenda, as well as eloquently highlighting the threats to wildlife. Bill Gates has used his prestige and reputation for hard-headedness to advocate practical policies for achieving net zero carbon emissions. And Greta Thunberg has energized the generation that will still be alive at the century's end.

Politicians plainly respond to public and media pressure, even if not to in-house advice. They will make wise long-term decisions – but only if they feel they won't lose votes thereby. As J.-C. Juncker said in another context: 'We know what to do, but we don't know how to get re-elected when we've done it.'[9]

Our policies need to take more account of the long term. We know how much we owe to the heritage left by our forebears. We surely have an obligation not to leave a depleted and dangerous world for future generations – to be 'good ancestors'. Today's children can expect to live to the end of the century. Policymakers should pay regard to their life chances, and indeed to the lives of those as yet unborn. One recommendation is that the UK Treasury formula (in the so-called Green Book), which sets the criteria for evaluating long-term public investments, needs revision. In calculating the value-for-money of long-term investment, it discounts the present value of most future benefits at 3.5 per cent per year until 2050.[10] This is surely too high in regard to ensuring mitigation of long-term threats.

At an international level, we will also need to take action. No single country wants, or has the power, to prevent risks such as pandemics or climate change on their own. Global risks require global cooperation. It is a false economy for the world not to explore potentially catastrophic scenarios, not to take action to minimize their probability and not to be prepared for future catastrophic risks. When crises hit, the costs can be trillions: sums of more than 20 trillion dollars are cited as the global cost of Covid-19, plus of course millions of deaths and the infection of hundreds of millions.[11] In that perspective a world investment of hundreds of billions of dollars in early planning and preparedness wouldn't have been disproportionate: it would plainly have significantly mitigated the pandemic's spread and impact.

We need to think globally, we need to think rationally, and we need to think long term. Though it has delivered so much, science remains an untapped resource. We not only need to consider how and what scientists should provide to governments, but how governments should provide for the scientific enterprise.

3.4 The scientific enterprise, national and international

Scientific knowledge is collective, public and international – it is, in principle, accessible to the entire world. But its benefits can only actually be captured by those who are educated and discerning enough – who are 'plugged in' to the research community. That's why it's in the interests of each country to maintain strong and broad expertise. As a UK citizen I'd find it dismaying if the standing of our universities and research were in jeopardy: once the tap has been turned off, it can't readily be turned on again. And I'm sure that scientists in any other country would feel the same about their universities.

This is not the first time that scientists have worried about their profession being in crisis. In 1831, Charles Babbage, the polymath and computer pioneer, wrote a book entitled *Reflections on the Decline of Science in England*. He deplores 'the gradual decline of mathematical, and . . . physical science, from the days of Newton to the present'. (His book is mainly, by the way, a diatribe against the futility and corruption of the Royal Society at that time. I'm an ex-President – so others must judge to what extent it has redeemed itself since the nineteenth century.)

There is also nothing new in the debate about whether science is being productively applied. To quote Babbage again:

In science, truths which are at one period remote from all useful application, become in the next age the bases of profound physical inquiries, and in the succeeding one . . . furnish ready and daily aid to the artist and the sailor . . . It is important to the country that abstract principles should be applied to practical use.

We now recognize that the so-called 'linear model' of innovation that Babbage implicitly assumed is naive. There is actually a two-way interaction between science and technology: science depends on improved instruments as well as underlying their design. Advances in my own field, for example, are due largely to better technology and computers. Armchair theory alone can achieve little. It has been said that there are two kinds of science: applied and not yet applied.

In his State of the Union Address in January 2011, President Obama urged the need for greater investment in science and technology. He recalled how the Apollo programme in the 1960s, a response to the Soviet Sputnik, had provided a broad impetus to technology and education, and asserted that his nation faced another 'Sputnik moment'. He offered a metaphor: 'You can't make an overweight aircraft more flight-worthy by removing an engine.' After the erratic tenure of Trump, this message is being renewed by Biden's administration. It's even more vital for the UK, which must, in recovering from the pandemic, rebalance its economy away from over-dependence on the financial sector, and acknowledge science and innovation as essential 'engines' for long-term prosperity and confronting global challenges.

In policy contexts the word 'science' is widely used to embrace technology. Even the most rarefied blue-skies researchers will surely be delighted if their work has a social or economic impact outside academia – even though they may not have the disposition or skills to develop these applications themselves. It is not always recognized how unpredictable, diffuse and long term such outcomes are. Even in targeted medical research, most new drugs take up to 20 years to develop. And the family tree of innovations in other fields stretches back even further in time, and is more diversely multi-branched. That's why, incidentally, the international effort that developed vaccines against Covid-19 with such unprecedented speed deserve special acclaim.[12]

The mechanisms for funding and organizing research are different across countries. The simplest system was in the old USSR where the Academy of Sciences, with its finances firmly under state control, had oversight of dozens of research institutes. In the UK, our academies – the Royal Society, together with the British Academy (covering the humanities), the Royal Academy of Engineering and the Academy of Medical Sciences – ensure that the UK has strong independent voices supporting research. In the US, the National Academies of Science, Engineering and Medicine perform a parallel role.

But, of course, the downside of their independence is that the budgets these Western academies directly control are trivially small compared to government funding – a very different set-up from Russian and East European academies in the Soviet era. That means that the health of the research culture in most Western countries depends on public bodies that channel taxpayers' funds in accordance with government priorities (at arm's length from government). In the US these are the National Science Foundation (NSF) and the National Institutes of Health (NIH); in the UK the analogues are the Research Councils. The allocation of these funds is decided by an array of subsidiary committees consisting of a mixture of officials and academic experts. We all grumble about the time-consuming bureau-cracy that this entails. But this is to some extent inevitable because most funds are distributed in small chunks, with all involved being concerned about fairness in allocating them when they are hugely oversubscribed. In Europe and North America, these public funds are supplemented by private philanthropy; and by companies, especially in high-tech and pharmaceuticals. (Such support is crucial, but of course donors are increasingly scrutinized for their probity, and their ideological or commercial bias.)

Despite lacking huge funds, academies have enough clout to engage effectively with governments, advising on priorities and policies. (The National Academy of Sciences has an obligation to do this – it was set up by Abraham Lincoln partly for this purpose – whereas the Royal Society remains a private foundation, and only more recently engaged formally with government.)

Of course, the epochal tectonic shift in the world's science stems from burgeoning growth in the Far East, in China above all. We are witnessing the terminal phase of the North Atlantic hegemony that has prevailed for four centuries. China's R&D spend has risen to a level that's now second only to the US – and in some key areas is now number one. China's technocratic leadership has astutely targeted its scientific investment on growth areas like genomics and AI. And its hybrid economy has allowed the growth of mega-companies that match those in the US and surpass the scale of any in Europe.

But there has – unsurprisingly in such a controlled nation as China – been anxiety about the power deployed by non-state conglomerates, despite their welcome con-tribution to advancing key technologies. Indeed in 2021 we saw attempts to rein in the power of Jack Ma (CEO of Alibaba) and other leading Chinese CEOs. Western observers view this with ambivalence; it's widely felt that their globe-spanning US counterparts should be subjected to similar measures, but it is unlikely these would be imposed effectually unless this were done with multinational agreement. There is concern about being dependent on Chinese technology in strategically important areas like nuclear power and the 5G internet. Whether a stand-offish policy is appropriate for Europe and the US will depend, among other issues, on whether the US or China prevails in the markets of Asia and Africa.

We don't know what the twenty-first-century counterparts of the electron, quantum theory, the double helix and the computer will be, nor where the great innovators of the future will get their formative training and inspiration. But one thing seems clear: nations will decline unless they can sustain efforts in discovery and innovation. Largely because of the resurgent Far East, the UK, as one example, risks becoming a small player. Moreover, departure from the EU ('Brexit') has plainly been damaging because it introduces friction into previously smooth trans-European collaborations. What is needed is a ten-year road map, offering hope that, after the Covid disruptions, science can share the fruits of the recovery that it will help to generate. Nations with strong scientific traditions can then contribute more than their pro rata share towards creating the technologies the world will need in order to surmount the challenges highlighted in chapter 1, and to achieve a sustainable and more equitable world by mid-century.

3.5 Some personal perspectives on academies and networks: from 'just in time' to 'just in case'

I have been engaged with the activities of the Royal Society,[13] the scientific academy of the UK and Commonwealth, since becoming a Fellow in 1979 – with two spells on the Council, and involvement with many of its multifarious (indeed excessively numerous) committees. The Society has a very broad remit, covering science itself, advice on government policies, and a strong international dimension. In 2005, I was elected to a five-year term as the Society's President, so I can offer an 'insider's' perspective – though of course this is doubtless a somewhat biased one.

The post of President is honorary, and therefore can only be part time for anyone who is neither retired nor independently wealthy. But there were many activities (fundraising, engagement with Fellows, 'representational' events, attendance at inter-academy meetings overseas, and so forth) where I felt the Society would have benefited from having a President with fewer other commitments. In fact, if I'd known that I was going to have this job, I wouldn't have taken on, a year previously, the Mastership of a Cambridge college; this took up a lot of time, despite being a non-executive role.

Many things remained undone during my Royal Society tenure, basically because the Society is relatively small in terms of senior full-time staff, and consequently limited in its expertise and impact. It is smaller (by most measures) than UK learned societies like the Institute of Physics and the Royal Society of Chemistry. It depends on the pro bono commitment of four other officers (supplemented of course by many other Fellows) and the work of a small number of senior people on the staff. I regretted this ineffectualness, particularly because there were several trends in UK government policy towards science and universities that made most of us anxious and where the Royal Society's voice could make a difference. Higher education policy was becoming a political football; a new ministry called DIUS (Department of Industry, Universities and Science) was set up – but dissolved two years later; the autonomy of the research councils was being eroded, and their problems were aggravated by ill-handled reorganization. Most important of all, it was crucial to ensure that science didn't fare too badly in the 'austerity' measures introduced by the Coalition government after the 2008 financial crisis, and was recognized as an investment in activities where the UK was currently strong, but where we were vulnerable in a world of mobile talent and fast-growing opportunities elsewhere, especially in East Asia. I think the Society indeed made some difference, through our reports, as well as influence on ministers. And my successors have confronted more serious issues: dealing with the scientific fallout from Brexit, and helping meet the scientific challenges of Covid-19 – and doubtless felt similarly frustrated.

I think what is needed to enhance scientific influence on the UK government is a federation of scientific societies, maybe convened by the Royal Society and the Royal Academy of Engineering. Such a federation, with a far greater aggregate membership than the Academies themselves, could claim more credibly to represent a nation's entire scientific community. An issue – which would then be in sharper focus – is the extent to which academies and learned societies should become advocacy groups for specific policies. Clearly they should offer assessments of scientific issues and lay out policy options; they should make recommendations within their range of expertise; they should offer views on the curricula of schools and colleges. But they should not adopt any collective stance that's too controversial, either through being overtly party-political, or because there's no consensus among experts. For instance, should we build more nuclear power stations? This is an issue where opinion in many countries is roughly equally split, both among people with genuine expertise, and among those with none. My line in the UK was that the Royal Society should not take a collective view on this, though I expressed my personal view in favour of R&D into improved nuclear reactors.

But what about issues when there is a strong consensus among experts but some 'dissidents'? This sharpened up in the context of climate policy. Our policy was that a collective Royal Society statement required endorsement by the Society's Council (the officers plus 18 elected members). There are around 1,000 UK-based Fellows altogether, among whom there will obviously be proponents of 'dissident' viewpoints, but these cannot expect to veto a statement. On this basis the Society endorsed the UK's 2008 Climate Change Act, which enshrined the goal of major cuts to CO_2 emissions, and was strengthened in 2018 to a goal of net zero by 2050.

Apart from climate policy, another issue that aroused controversy (though fortunately peripheral to the Society's main agenda) stemmed from a vocal faction of 'new atheists' – best described, I think, as small-time Bertrand Russells, as there was little in their views that he hadn't expressed more eloquently decades earlier. My line was that the Society should be a secular organization but need not be anti-religious. Of course we should oppose, as Darwin did, views manifestly in conflict with the evidence, such as creationism. But we should strive for peaceful coexistence with mainstream religions, which number many excellent scientists among their adherents. This tolerant view would probably have resonated with Darwin himself, who wrote: 'The whole subject is too profound for the human intellect. A dog might as well speculate on the mind of Newton. Let each man hope and believe as he can.' If teachers tell young people that they can't have both God and Darwinism, many will choose to stick with their religion, and be lost to science. My own perspective is that if we learn anything from science it is that even something as basic as an atom is quite hard to understand. This should induce scepticism about any claim to have achieved more than a very incomplete and metaphorical insight into any profound aspect of our existence. But this need not prevent us from appreciating the cultural traditions, rituals and aesthetic accretions of religion, and its emphasis on common humanity in a world where so much divides us.

There was some flummery at the Royal Society in my final year as President, 2010, when the Society's 350th anniversary was celebrated. A range of media, regional and international events, and an exhibition on the South Bank, raised the Society's profile and helped to promote science generally. There was even a commemorative service in St Paul's Cathedral (though the 'new atheists' might have been gratified that halfway through it, the fire alarms went off and we had to evacuate the building!). The centrepiece of the anniversary was a celebration in the Royal Festival Hall, attended by the Queen, five other 'royals' and a 2,000-strong international audience.[14]

The BBC marked the anniversary by dubbing 2010 their 'year of science', with an increase (albeit superimposed on a falling trend) in its science coverage. As part of this initiative, I was invited to give the Reith Lectures, which subsequently appeared, in expanded form, in a small book called *From Here to Infinity: Scientific Horizons*. While it would seem a doddle to present four half-hour radio talks, the format of the lectures made it rather more demanding as they were arranged as formal public lectures, each in a different location.

My five-year stint as President of the Royal Society had involved a fascinating variety of events and contacts, and was punctuated by no more than a few irritations. I handed over to the biologist Paul Nurse – an obvious and popular choice who 'ticked all the boxes', though as it turned out he had far more demanding conflicting commitments than me, as the founder-director of the Francis Crick Institute.

The Royal Society presidency gave me inside knowledge of one academy – but of course there were many issues where international dialogue was needed – for instance the regulation of biological experiments, and collective campaigning in advance of G7 and G20 meetings dealing with climate, environment, energy or regulations. These were addressed at inter-academy discussions. One stumbling-block was that, though more than a hundred nations have 'academies', these vary in their structure and scope. The US National Academy[15] is a powerful body. And in Europe several countries apart from the UK, including Germany, Sweden and Holland, have effective academies. In contrast, other national academies are smaller and essentially just honorific, and in many cases geriatric (a recent president of the Japan Academy was 84 years old on appointment); these institutions consequently can't plausibly influence their country's policies, nor speak on behalf of their entire scientific community. The world's academies are therefore collectively less than the sum of their parts, and have not been as effective as federations of learned societies covering specific subjects. A welcome innovation was the formation, in 2018, of the International Science Council,[16] which gathers together around 200 organizations: both national academies and international subject-specific organizations. And, of course, 'concerned scientists' can exert leverage through engagement with international charities that focus on global health, disaster relief or conservation.

National academies like the Royal Society each have individual memberships – indeed it's fair to say that choosing them is for many members the most absorbing activity. These elections can have the same downsides that prizes do (see section 4.5), though for smaller stakes. But there's a vulnerability here. Members have to be elected – but before that they must be nominated by existing members. The scrutiny of nominees is generally thorough. But many excellent people aren't nominated, because this requires an initiative (and some effort) from one or more existing members. This system is fine if most members are conscientious and feel an obligation, or if there are proactive searches for candidates. But there's a risk of a downward spiral if sustaining the academy becomes a lower priority for increasingly busy academics – and academy members lose interest in making nominations. It's a good benchmark that, to have credibility, an actual academy should be stronger than a 'virtual academy' that could be constructed from eligible non-members. I'm marginally reassured that the Royal Society can claim, at least for now, to remain well above that (rather low) threshold.

Perhaps one of my more incongruous involvements (at least at first sight) has been with a rather unusual academy – one with an international membership. This is the Pontifical Academy of Sciences,[17] established by Pope Pius XI in 1936. Its members, about 70 in number, embrace all faiths (and none). It holds regular plenary meetings. But more valuable are its specialized 'Study Weeks'. I first attended one of these, on galactic nuclei when I was a postdoc. I have continued my association, becoming a member 25 years ago and serving on its Council. There are obvious tensions and constraints operating within the Vatican; but an in-house Academy can have a positive influence – indeed substantial leverage. Perhaps its most influential and worthwhile meeting took place in May 2014, as a joint effort of the Pontifical Academies of Science and of Social Science. It was on the theme of environment, biodiversity and climate. Co-organizers were economist Partha Dasgupta and climatologist V. Ramanathan (two 'lapsed Hindus'!); speakers included the economists Joe Stiglitz and Jeffrey Sachs, along with environmental and climate scientists of that stature from all over the world. What made this meeting especially worthwhile was that follow-up within the Vatican led to the encyclical *Laudato si'*,[18] issued in June 2015, which was clearly influenced by the input from the Papal Academies. This document, boosted by the Pope's visit to the UN in September, where he got a standing ovation, was influential on voters and leaders in Latin America, Africa and East Asia – and eased the path towards consensus at the important COP conference in Paris in December 2015. More recent academy meetings – on extinction, ethics of transplant surgery, AI and human trafficking – have been less impactful but still worthwhile, gaining added traction though the Vatican's imprimatur.

In 2005, I became a member of the UK's House of Lords in the category of 'people's peers'. This process involved being nominated, and then, if shortlisted, being interviewed by a panel. After being interviewed, nothing happened for more than a year, but I then was told that I had been selected for appointment. It was important that peers in this category should be crossbenchers, with no party affiliation. I had been a Labour Party member for 30 years and had been active, especially, in the Kinnock campaign in 1992. But I had been less active subsequently, describing myself as 'Old Labour' and not 'New Labour'; so I felt more at ease on the crossbenches. Moreover, the 'working peers' on the Labour benches are subject to the Whips and therefore have to make a greater time commitment than crossbenchers – more than I felt I could do. Most new peers, however, enter via a different route: nomination by the Prime Minister or party leaders. Numerous such appointments in recent years have been criticized as rewards for donors or cronyism: there's a widespread view that reforms are needed, and that the House should be reduced in size. But membership remains a privilege, even though perhaps a diminishing honour.

Although many members work intensively in amending legislation (a session in February 2022 continued until 3.30am), I have honestly been guiltily inactive in the Lords debating chamber – speaking only on themes where I think I can contribute usefully (or feel strongly, as in the 'assisted dying' debates). I've spoken on university issues (fair access and research); in particular, I was a vocal opponent of the setting-up of UK Research and Innovation (UKRI) – an unwieldy conglomerate which over-centralized research funding under one supremo – rather as though, in the US, the NSF, the NIH, the Defense Advanced Research Projects Agency (DARPA) and the Council for the Humanities all had a single line manager. At the time of writing, it is too early to judge whether these forebodings are fulfilled – this is a context where I obviously hope I was wrong.

Reports by parliamentary committees (and those by similar committees in the US Congress) have a distinctive value when they thoroughly address long-term issues and recommend legislation where cross-party consensus is necessary. I've been appointed twice to four-year stints on the Lords Select Committee on Science and Technology (and have served also on committees that scrutinized EU affairs). But the most worthwhile has been membership of a 'Special Inquiry' on 'risk assessment and risk planning' that met throughout 2021 and reported at the end of that year.[18] In fact I had instigated this inquiry, with the support of more than 20 other peers, because the UK's initially confused response to Covid-19 had demonstrated how ill-prepared the country was to cope with emergencies. Pandemics are unpredictable, but not hugely improbable; moreover, we need to be mindful of a range of other catastrophes – some global; others regional or local, like cyber attacks, grid breakdowns, radiation releases, failures of ancient infrastructure, or floods. They're rare enough to be easily ignored – even though the worst of them could be so devastating that one occurrence is one too many. It's a wise mantra that 'the unfamiliar is not the same as the improbable'.

The committee benefited from an excellent chairman, James Arbuthnot, and a strong cross-party membership including seven former ministers and two former secretaries of defence. Among our 85 invited witnesses were experts from several other countries, able to share their own experiences with us, as well as many from central and local government and from military, business and scientific backgrounds.

The UK government maintains a National Risk Register. This is intended to be a comprehensive document, classifying threats in terms of likelihood and severity, and offering guidance to the public and private sector on appropriate preparations and precautions. Many witnesses bemoaned its current inadequacies (and the undue secrecy in which it was prepared). Famously, it had highlighted the risk of influenza, but rated any other pandemic as unlikely to cause more than a hundred deaths. Moreover, it didn't take account of the likelihood that disasters could cascade between sectors – for instance in many countries the pandemic had massive effects on the education system. We proposed an Office for Preparedness and Resilience, scrutinized by a Parliamentary Select Committee and an annual debate in Parliament.

On a national level we need to rebalance the trade-off between resilience and efficiency. For example, if manufacturers depend on supply chains spanning the world – and 'just in time' delivery – they're vulnerable to a break of one link in one chain. We should shift focus from 'just in time' to 'just in case'. And – to take another example – it may be 'efficient' to routinely have high occupancy of intensive care beds in hospitals, but this is imprudent as it leaves too little spare capacity for emergencies.

Of course, mitigation of global mega-risks is an international challenge. For instance, there's a need to render the internet more resilient – and to ensure that the dozens of top-security (level-four) biological labs worldwide are indeed safe; and that the WHO has the resources to identify emergent viruses quickly. Finally, our Committee's message was that our interconnected high-tech world is vulnerable. Without more foresight we'll face a bumpy ride through coming decades.

A nation needs to prepare for potential disasters; it should also ensure that its physical infrastructure and 'human capital' are geared towards enhancing national wellbeing in a fast-changing world where science not only helps us cope with threats, but is crucial for economic and social advances. In chapter 4, I address how these aims can be achieved through optimizing education and the research environment, and incentivizing scientific careers. More generally, I believe it is essential to boost our broader understanding of what science is and what it can achieve – to make it the public culture we so urgently require.

Chapter 4

Getting the Best from Science

4.1 Optimizing scientific creativity

I'm fortunate to know many of today's leading scientists – those who have achieved conceptual or experimental breakthroughs. They are individualists, spanning a range of personalities and expertise, but most have some things in common. Apart from a few whose discoveries were serendipitous, they staked their careers on a specific line of research – and they chose well. The path they took was unpredictable, and often the intellectual pay-off was long in coming. The two Russian winners of the 2010 Nobel Physics Prize, Andrei Geim and Konstantin Novoselov, are exemplars of 'brilliant loners'. They made the unexpected discovery that carbon atoms could form a lattice just one atom thick – a new material, graphene, with exceptional strength and many potential uses.[1] In contrast to those who work in particle physics or space science, they didn't need major equipment – the clinching experiments involved strips of Scotch tape. But these men staked several years of their lives, and their reputation, on their quest. And the University of Manchester offered the security and intellectual freedom they needed.

We can't confidently predict how, when or whether a specific research project will pay off intellectually – still less when (if ever) its applications will offer social or economic benefits. But successes are favoured by a nurturing environment. In the words of a former Royal Society President, biochemist Aaron Klug:

> The major insights in science come from people who have the patience to develop an intimate understanding of a problem, who have the space and the freedom to take professional risks and who know how to make creative use of the surprises that they encounter when they do so. These are the people who make the enduring difference. These are the people whom we must nurture wherever we find them.

Confidence and high morale drive creativity, innovation and risk-taking – whether in science, the arts or entrepreneurial activity. And even when not overtly collaborative, no achievement should be credited solely to an individual: those who foster the intellectual and social environment are crucial too.

Exploring the applications of graphene demanded greater resources than its discovery did. It's an extraordinary material. Thin sheets have exceptional strength, and (as a genuine surprise) two sheets, when aligned in a particular way, behave like an electric superconductor. Some of these properties can be explored and exploited by team efforts in universities and in government-supported research institutions – indeed, the European Union earmarked a billion euros towards this goal. But the main pay-off, via large-scale manufacturing, will require the resources and expertise of high-tech companies – working in the style that has brought us smartphones, new drugs and other triumphs of advanced technology.

In research, second-rate work counts for very little – international excellence is all. Researchers themselves have the best expertise for judging what topics hold promise of yielding exciting science, and the strongest motive for choosing an area in which they'll have an impact. The difference in pay-off between the very best research and the merely good is, by any measure, thousands of per cent. So what is most crucial in giving taxpayers enhanced value for money isn't the few per cent of savings that might be made by improving efficiency in the 'office management' sense. It's maximizing the chance of the big breakthroughs by attracting top talent, backing the judgement of those with the best credentials, and supporting them appropriately.

Research universities won't stay internationally competitive unless they can attract and nurture such people. Scientists are of course accountable to their funders, but it's essential that those with a good track record can follow their own judgement rather than being constrained by narrow external targets.

A perennial tension within funding bodies is between the support of people, and the support of specific projects. The latter option – sadly becoming more dominant – is administratively tidier, and allows the funder to demand 'quarterly reports' of progress and keep track of steps towards a declared target. But history shows that it's often free-wheeling inquiry that leads to the biggest advances. In lively research groups, it's exhilarating when coffee-time conversations toss out new ideas and occasion debate on the latest discoveries. The best institutions all foster such an atmosphere, and I'm lucky to have worked in one of them. But even in this kind of privileged environment, my younger colleagues seem ever more preoccupied with grant cuts, proposal writing, job security and suchlike. Prospects of breakthroughs will plummet if such concerns prey unduly on the minds of even the very best young researchers. Not just in the UK, but in the EU, in the US and elsewhere, bodies that allocate public funds for science and education focus on ever more detailed 'performance indicators' to quantify the 'output'. This has the best of intentions – to raise standards, improve accountability and enhance the chances of beneficial spin-off. But its actual consequences are often the reverse: to impede the best professional practice. University research offers incontrovertible benefits to the economy and to society. But there's a danger that a focus on what can be measured, and can feature in league tables, may distort policy, to the detriment of longer-term benefits.

Research isn't a zero-sum game. It would indeed be a stimulus to the UK if there were more top-tier research universities in the rest of Europe – incentivizing greater mobility and opportunity. Europe collectively could then offer a stronger counter-attraction to North America (and to China or Singapore) as a destination of choice for mobile talent. Indeed, there has been a beneficial change since the 1970s. At that time, young UK scientists tended to meet their European counterparts because all went on fellowships to the US. Now there is more interchange among European countries – and, of course, with China, India, Japan, Korea and Singapore. Sadly, this risks a serious setback due to Brexit – its reality and (even more) the perception.

In the so-called 'big sciences', there's long been well-managed European collaboration, and European consortia have achieved real excellence. CERN in Geneva, home of the Large Hadron Collider (see section 2.5), is destined to remain the world's leading laboratory in particle physics for at least the next decade. Europe's space programme has never matched the overall scale of that of the United States – which spends vastly more on defence projects and on human spaceflight. But Europe could rise beyond parity and gain an ascendency in space science (including environmental monitoring) if it focused on robotics, leaving NASA to squander much of its far larger budget on a programme of human spaceflight that is neither useful nor inspiring.

Grand flagship projects in particle physics, space and astronomy aren't, of course, typical of scientific research overall. But they're good portents of long-term funding, and they show that Europe can fully match the US by optimally developing a collaborative research community. And it's fortunate for Europe that CERN, the European Space Agency and the European Southern Observatory (which is now building the world's largest optical telescope) were governed by separate protocols, so Brexit didn't lead to the UK's expulsion from them.[2]

It is in the interests of all advanced nations to foster the expertise needed to meet high-tech challenges – for instance, the capability to create new vaccines, and to develop technologies that accelerate the transition to clean energy. The research itself may be concentrated in universities, industry or government laboratories: not all nations choose the same balance between these three. But a crucial prerequisite is an effective education system and a national ethos that encourages and incentivizes scientific achievements. And it is crucial to remember that the development and deployment of any scientific innovation for public benefit depends on a wide range of technical, craft and 'human' skills – and will be scuppered if there's no improvement in the training and education of those young people who don't pursue 'academic' higher education. So the pre-18 stages of education of all young people – not just an academic elite – are crucial to a successful and prosperous nation.

4.2 Educating scientists: an international perspective

Education at all levels is prioritized in China – and in Taiwan, South Korea, Singapore and other countries of the Far East – with a focus on fulfilling the aspirations of their fast-developing economies. These countries are already ahead of the West in key aspects of school-level education. Attainment levels of UK and US pupils are poor by international standards[3] – a depressing augury for the West's future. There are not enough good science teachers to ensure that every pupil gets exposed to one. Young children display enthusiasm and curiosity – often focused on dinosaurs and the cosmos (blazingly irrelevant to their lives, but fascinating). But all too often they're denied the inspirational teaching that could build on this enthusiasm. In consequence, a substantial fraction of our young people are deprived of the chance to qualify for the most competitive university courses – and may even be 'turned off' science completely. Despite many initiatives, and some positive trends, substantial improvements will be slow. In the short run there are three things that can be done: ensure that conditions are good enough to retain excellent school-teachers, and that their pay level is appropriate for practitioners of a serious profession; encourage mature individuals to move into teaching from a career in research, industry or the armed forces; and, thirdly, make better use of the web and distance learning.

At the university level, the West remains better-placed – and the UK ranks higher. British universities have problems, but they compare well with those in the larger countries of mainland Europe. One should be properly cynical about the spurious precision of the various 'league tables', but it counts for something that the UK is the only country outside the US with several universities ranked in the 'premier league'.

But the UK's ability to attract and retain mobile academic talent is now at risk. Even to retain its international competitiveness despite the setback of Brexit, the nation must raise its game. There's now an international market for the best students as well: they are academic assets, and a long-term investment in international relations. After they graduate, they'll feed into all walks of life, networked worldwide, ready to seize the best ideas from anywhere and run with them. The focus is now on those from China: they comprised (in 2021) about 10 per cent of total enrolment in the UK (and 20 per cent in at least one major university).

Concerns have sometimes been voiced about accepting students from countries that are deemed potentially hostile. I think these concerns are overplayed. The quality and volume of Chinese research is now so high that we will lose as much as we gain by inhibiting exchanges. Moreover – and this is admittedly more controversial – I think we should maintain contacts with, for instance, Iran. There were in the past, for instance, refusals to admit Iranian students for courses such as nuclear physics. But these students will learn nuclear physics somewhere, whatever barriers the US and UK impose, so it's surely better that they should in the process make contacts with us, and retain them – decreasing the chance that clandestine programmes can proceed without someone in our country becoming aware of them. There was an enlightening instance of such benefits during the multinational talks in 2015 aiming to restrain Iran's development of nuclear weapons. The Iranian minister for atomic energy, Ali Akbar Salehi, asked the US negotiating team to include Ernie Monoz – a distinguished physicist who was then the US Secretary for Energy. These two men knew and trusted each other through having studied at MIT at the same time.

And what about the migration of trained graduates? The phrase 'brain circulation', rather than 'brain drain', is now a more apt description of what's happening among Western nations, and with advanced economies like China, Singapore, Taiwan and South Korea. Migrants can now, unlike in earlier decades, retain contact with their homeland: communications are always open; travel is far easier and cheaper. Three of the greatest US-based companies now have Indian CEOs: Microsoft (Satya Nadella), Google/ Alphabet (Sundar Pichai) and IBM (Arvind Krishna).

But this mobility offers little consolation to the least developed countries: they face daunting challenges in retaining their all-too-few highly trained people, and even more in attracting them back. Those of us in the developed world should surely be uneasy and feel an obligation to redress this loss. Africa's predicament is the worst. Around

half of its health workers want to leave and their departure can be ill afforded; it's doubly tragic if, after moving
to a developed country, they find they're not accredited,
and doctors become cab drivers. It's just as bad in agricultural science, engineering and all the other specialties
that African countries require if they are to develop their
potential. (And the nations of the North, responsible for
most of the world's anthropogenic CO_2 emissions, surely
have a special responsibility to aid with measures to cope
with climate change.)

The poorest countries need to engage their diaspora communities, encouraging those with expertise to at least make
regular visits back home. But wealthier nations should take
some responsibility too. A cost-effective form of aid would
be to establish, in Africa and elsewhere, 'centres of excellence' – with strong international links – where ambitious
scientists could work in less dispiriting conditions, perhaps
via linkages with foreign experts. They could then fulfil
their potential without emigrating, and strengthen tertiary
education in their home country – as well as working with
other countries on the challenges of clean energy and intensive agriculture, on which (as mentioned in chapter 1) their
future depends. It would seem equitable, in the meantime,
that for each skilled person 'drained' to the developed world,
the receiving country should feed back sufficient resources
to train two more.

There have been some modest-scale efforts to support
scientists at risk of intellectual isolation. For instance, the
International Centre for Theoretical Physics (ICTP) was set
up in 1964 by the brilliant and inspirational Pakistani physicist Abdus Salam as a place where physicists from what
was then called the third world could go to recharge their

intellectual batteries and stay in touch. Its fine buildings in Miramare, near Trieste, now host meetings and schools over a range of physical and environmental sciences, though physics remains the core. Salam also founded the Third World Academy of Sciences (TWAS). Although it has kept the same acronym, its full title (more 'PC') is now 'Academy of Sciences for the Developing World'. This too is based in Trieste and aims to promote international contacts among scientists in developing countries.[4]

4.3 Attracting and supporting scientific talent

Of course only a fraction of young scientists will become teachers or academic researchers. But it's crucial for the next generation that enough should choose this career. The traditional 'compact' which attracts scientists to academia is that in return for their teaching, they can devote part of their time to research in fields of their own choice, and have reasonable prospects of the necessary support. This has manifestly paid off in places like Harvard, Berkeley and Stanford – each an immense asset to the US. The UK must not jeopardize its counterparts of these great institutions by putting this compact under threat.

It would be bad news if talent were not attracted towards academia at all – if young people developed a negative perception of the profession. Charles Babbage (the nineteenth-century innovator quoted earlier) had something to say about this too:

> Let us now look at the prospects of a young man impelled to devote himself to the abstruser sciences *[no women men-*

tioned, even though Ada Lovelace was his most discerning collaborator!]. What are his prospects? Can even the glowing pencil of enthusiasm add colour to the blank before him? . . . What can he reply to the entreaties of his friends, to betake himself to some business, or to choose some profession like the law in which his talents may produce for him their fair reward?

We have these tensions today. In my Cambridge college I asked a group of final-year engineering students what their career plans were. All but one were heading for finance or management consultancy. At a time when the UK, in particular, needs to reduce its dependence on the financial sector, and rebalance towards high-tech manufacturing and services, the sciences must attract a share of those who have a choice of career paths and who are mindful that the City of London still offers Himalayan salaries, if no longer such high esteem. We're often reminded in the UK of the global financial role of the square mile in the City of London. But is that really a match for the impact on today's world of the ideas that gestated in the square mile where a historic university like Cambridge or Oxford stands? We shouldn't be bashful about what our forebears here achieved – it should be celebrated, and should spur us to sustain this role. It's crucial for the world that enough of its brightest young people, in all countries, should perceive science and technology as offering attractive opportunities.

Some people will become academics come what may – the nerdish element (of which I am one). But a world-class university system cannot survive just on these. It must attract a share of young people with flexible talent, who are savvy about their options and ambitious to achieve

something distinctive by their thirties. They may associ-
ate academia with uncertain prospects and undue financial
sacrifices.

When I wrote an op-ed on this theme, the newspaper's
editor gave it the headline 'Why I'm glad I'm not a young
academic.' And I am – to the extent that I feel fortunate to
have started when academia was less crowded, and when
the profession was still growing fast, riding on the overall
expansion of higher education during the 1960s. In university faculties, the young then outnumbered the old, and
the old then retired in their sixties, thereby opening up new
'slots'. It was reasonable to aspire to lead a group while still
quite young. That's far less true now. In the US, the average
age of those receiving their first grant from the NIH has
now risen to 44.[5]

A further off-putting trend endemic in many countries
is the pervasive 'audit culture' and the deployment of ever
more detailed performance indicators to quantify our
outputs. When academics extol free-wheeling research,
they risk being accused of an ivory tower arrogance that
disregards their obligations to the public. But they should
counter such allegations. Choices of research project are
anything but frivolous: what is at stake is a big chunk of
our lives, and our professional reputation; not just money
is being staked.

A system of 'Research Assessment' was introduced in
the UK in the 1980s with a benign purpose – to allocate
more public money per student to departments doing good
research – so that those departments could improve staff–
student ratios and give their faculty extra time for research.
That's still the alleged aim. But it's now hugely burdensome.
In making appointments, universities focus on what will

score best in the next research assessment. Moreover, this pressure gives two perverse incentives to young academics – to shun high-risk research; and to downplay their teaching. The most effective way to improve university teaching would be to make research assessment less vexatious. It is by enabling committed individuals to back their own judgement that funding agencies will best sustain high-quality and cost-effective universities.

In addition to this, the current incentive system underrates something that is surely part of an academic's remit: broad learning and scholarship. The Robbins Report[6] – a manifesto for UK university expansion in the 1960s – stated that an academic had three duties: teaching, research, and 'reflective enquiry'. 'Reflective enquiry' is now being squeezed out. It is important for its own sake, as well as for the way it enriches both teaching and research.

There is now, as described in section 4.6, a welcome involvement of 'citizen scientists' – hobbyists enabled by the internet to participate in data-gathering and other tasks that help to advance research. But a further likely trend in coming decades, and a benign one, may be the full-time involvement in frontier research of professional-level independent scientists, some with substantial resources earned in high-tech businesses, who could, free of institutional constraints, enrich the intellectual landscape and offer some newly original perspectives of a kind that might otherwise be stifled – just as Darwin, Rayleigh and other wealthy scholars did in the nineteenth century.

4.4 Where is research best done – and best exploited?

One of the things that I've tried to highlight in this book is that the potential of the scientific endeavour should not be underestimated or viewed too narrowly. It needs ambitious and stable funding, in a variety of institutions. Many of us are used to the idea that research is mainly concentrated in universities; this is the system in the US and the UK. But it doesn't prevail everywhere. Although the research university was pioneered in the early nineteenth century by Humboldt in Germany, most of that nation's best researchers are now in Max Planck Institutes; in France they're civil servants in CNRS institutions. So the kind of academic career that mixes teaching and research is an Anglo Saxon model, though it's now widely adopted in the Far East.

But universities aren't necessarily the most propitious environments for research projects that demand intense and sustained effort. Dedicated laboratories that can offer long-term support are in some contexts preferable (though there is a downside insofar as they remove talented researchers from contact with students). Indeed, one reason why the UK has developed special strength in biomedical sciences stems from the existence of laboratories allowing full-time, long-term research that may be harder to do in today's universities. Moreover, the UK government's funding is massively supplemented by the Wellcome Trust, the cancer charities, and a strong pharmaceutical industry. In the US, likewise, funding from the NIH is supplemented by the Howard Hughes Foundation and mega-scale private philanthropy. And there have been examples in the past (most famously the Bell Telephone Laboratory in Murray

Hill, New Jersey) where industrial labs have germinated exciting blue-skies research.[7]

To ensure effective exploitation of new discoveries, research institutions must be complemented by organizations, in the public or private sector, that can offer adequate manufacturing capability when it's needed. This fortunate concatenation certainly proved its worth in the recent pandemic – in vaccine development and production, analysing virus variants, etc. It's imperative likewise that nations should foster expertise in energy, climate and the cybersphere – indeed in all the fields of natural and social science needed to tackle the global challenges highlighted in chapter 1 – and that research and development should be accommodated in a range of institutions of complementary strength, to facilitate blue-skies speculation, long-term programmes and a rapid response to emergencies.

Research universities benefit the economy partly through direct knowledge transfer from university labs to industry – but perhaps even more through the quality of the students they feed into all walks of life. But they offer an indirect benefit that's harder to quantify but could be just as important. They're a source of independent expertise, of people who are plugged into new ideas emanating from anywhere in the world. Moreover, such universities, harbouring the best research teams, have an advantage over free-standing labs in engendering direct knowledge transfer because they are multidisciplinary. And, as mentioned earlier, they should accept an obligation to provide advice to government on issues where the natural (and social) sciences are relevant.

We're all familiar with the symbiosis between great universities and a surrounding cluster of high-tech enterprises in the US: most spectacularly between Palo Alto and Stanford; and between Boston and Harvard/MIT. But there are parallel developments on the other side of the Atlantic. In Cambridge[8] and Oxford, for instance, dynamic high-tech communities have grown up around the universities: they offer a supportive environment where failure can be a step towards future success – it doesn't require personal upheaval because of the concentration of opportunities. These cities have become, in the words of Charles Cotton, 'low risk places to do high risk things': they can claim to host not only world-leading research universities, but also to be leading European concentrations of high-tech companies – ranging from research centres for global companies to start-ups. A genuinely benign symbiosis has developed between academia and dynamic high-tech communities – stimulated by personal friendships, and by mobility between the two sectors.

I would mention just two members of my Cambridge college who have displayed impressive business-savvy credentials as well as being top-ranked academic researchers. Greg Winter, Nobel Prize winner and my successor as Master of Trinity, was the inventor, via his start-up called Cambridge Antibody Technology, of Humeira, for some years the world's best-selling drug. And Shankhar Balasubramanian co-invented a technique for gene sequencing that was implemented by his start-up, called Solexa – but Solexa was taken over for about £600 million by a US company, Illumina, which has multiplied its value almost 100-fold. What's sad, and in many cases damaging to the UK, is that highly promising start-ups are too often bought out or taken over prematurely, usually by US companies. It is only partial consolation, though welcome, that several of the mega-companies, including

Microsoft, Google and Samsung, have set up research labs in Cambridge.

But Europe still lags in nurturing start-ups. Venture-capital funding remains far tighter than in the US. A measure is the number of new 'unicorns'[9] (start-ups capitalized at more than a billion dollars): the numbers created within Europe in 2021 were 27 in the UK, 15 in Germany, 8 in France, and 5 in Switzerland. But there's a long way to go; the number for the US is 288. But the entrepreneurial culture associated with the US West Coast has genuinely taken root in the UK, which ranks high in the academic impact of its research – in fields from biotech through to space science. But the UK doesn't yet harness this expertise optimally for commercial and social benefit, and there is rich potential in doing so: at a time of fast-moving change, it is handicapped because the non-governmental investment in R&D is lower than in other comparable nations. The fraction of the UK's GDP spent on research is 1.7 per cent – compared with 2.2 in France, 3.2 in Germany and 4.3 in South Korea. The OECD average is 2.5.

4.5 Why prizes can do more harm than good

What motivates scientists? Despite exceptional cases, like those I've just mentioned, where outstanding scientists have themselves spearheaded remunerative spin-offs from their discoveries, for most of those in academia the prime aspiration isn't to become wealthy. It's to advance scientific understanding, and thereby gain intrinsic satisfaction, earn the esteem of one's peers – and promotion within the academic system. There is rightly a growing focus on the fairness of such systems – and a special need to ensure greater diversity. It's an obvious 'own goal' if talent isn't being fostered and encouraged in our entire community. Everyone's achievements are dependent on a supportive infrastructure – and it's unhealthy if the system encourages younger colleagues to feel they're competing with each other for jobs and security. That's why I think we should be ambivalent about high-profile awards. Just as the pay inequalities in the private sector are grotesquely disproportionate, so is the focus of public acclaim on a lucky few scientists. This, at least, is my view, for reasons that I'll try to offer here.

Each year, the news is punctuated by award ceremonies – the 'Oscars', the Booker Prize for novels written in English, the Nobel Prizes, and the Turner Prize for visual arts. But are these prizes a 'good thing'?

Doubters were especially vocal in 2019. The Booker Prize had no outright winner: it was – contrary to precedent – shared between two novelists. More remarkably, the Turner Prize judges acceded to the finalists' request that all four of them should share the winnings. These decisions attracted flak: some thought that the judges were failing in their job. But of course assessment of artistic merit is inherently subjective and engenders dissension and debate. The safer (albeit duller) option is a vote by a wider group – for instance the several thousand authorized to vote annually for the winner of the 'best picture' Oscar.

It's different at the Olympics. Nobody suggests that the gold medal for the 100 metres should be given jointly to all finalists: the criteria for winning are in that case quite clear. (But in the 2020 Tokyo Olympics there was a tie in the high jump. Rather than settling a winner by a 'jump off', the two contestants, from Qatar and Italy, chose to share the medal – a decision widely applauded by spectators.)

What about prizes for scientists? Outsiders might guess that in these cases objectivity should reign and controversies be as absent as in athletic competitions. But that's not the reality. It's easy to agree on what scientific advances in any particular field of science are important (though there may of course be dissension about the relative status of different fields). But what's not so easy is to apportion credit for specific advances. A Turner Prize winner's creation may be ephemeral but it's their own. If they hadn't made that particular artwork, nobody else would have done so. But in science, if 'A' didn't make a particular discovery, then sooner or later (and usually sooner) 'B' would have done so. Moreover, each advance builds on the work of others. No scientist's achievements are really solo, any more than a goal-scorer's triumph in football is independent of the other players on the field (and the manager off the field too).

The Nobel Prizes cover some (but not all) fields of science, and of course have a distinguished lineage and great prestige. There's consequently been frequent controversy about the choice of winners. The Nobel Committee's refusal to make an award to more than three people has led to manifest injustices, and given a misleading impression of how 'big science' actually advances. Sometimes a discovery involves the cooperation of a large group. The 2017 Nobel for physics recognized a discovery, LIGO's detection of gravitational waves (described in section 2.5) – that was reported in a 1,000-author paper – but just three of these authors were rewarded. And the 2015 award, for studies of neutrinos using large-scale underground equipment, was given to the heads of two groups, one in Japan and one in

Canada. In contrast, the Breakthrough Prize for this same discovery explicitly recognized all members of five groups. In 2011 the Nobel physics prize went to astronomers for the remarkable discovery of 'dark energy' latent in empty space that causes the expansion of the universe to accelerate. This research, which I followed closely, was accomplished by two independent teams, each with up to 20 members. The award went to three people – even though others in the team made distinctive contributions, and several had track records as fully distinguished as those of the awardees. It was for that reason less satisfactory than the LIGO case, when there would have been a consensus among the team that the three individuals singled out had especially outstanding and prolonged records of achievement.

Even if a discovery isn't explicitly a team effort, several people may have separately researched the same topic and reached the finishing-line almost simultaneously. For instance, the Higgs boson was an idea that emerged in the 1960s as a 'capstone' of the standard model of particle physics: six people were generally agreed to have played key roles in predicting its existence (building on the work of still more, of course). Of these six, the one with the strongest and most sustained lifetime achievement, the late Tom Kibble from Imperial College, did not receive a share of the Nobel when the particle was discovered (nor did any of the thousand-strong team who conducted the huge experiment that actually discovered it, nearly 50 years after it was predicted).

Nobel prizes exclude huge tracts of science. Famously, mathematics has never been included. But other vibrant new fields are left out too; the environmental sciences – space, oceans, climate and ecology – aren't covered. Nor are computing, robotics and artificial intelligence. The Nobels currently distort the public perception of what sciences are important; they also, by failure adequately to acknowledge collaborative and parallel work, give a misleading impression of how science is done.

These flaws and gaps are being partially remedied by philanthropists who have established new prizes – some promoted with a razzmatazz that matches the Nobels, and with even bigger jackpots. Among them, for instance, are the Breakthrough prizes set up by the US/Russian billionaire Yuri Milner (which have been awarded to groups, such as the 1,000 experimenters at CERN who discovered the Higgs particle), and the million-dollar Berggruen prize for philosophy (given in separate years to three widely admired 'public intellectuals': Martha Nussbaum, Onora O'Neill and Ruth Bader Ginsberg). Overall, major awards now offer a better balance across the map of learning – both sciences and the humanities. Some awards offer substantial prestige but minimal prize money; for others, the reverse is the case.

Some argue that we should welcome the existence of high-profile mega-awards that elevate a few intellectuals to a transient celebrity status. But there is a downside: Nobel winners' opinions are sought by the press, and accorded undue respect. Even the best scientists (and artists) generally have narrow expertise; their views on broader topics carry no special weight. Some of the greatest scientists become an embarrassment if given a public platform. It's possible to find a laureate to support almost any cause, however eccentric – and some exploit their status. Laureates aren't necessarily exceptional intellects: some of the greatest discoveries have been made (serendipitously) by people who wouldn't claim any intellectual superiority to the average university professor.

So more of us are coming to query the societal benefit of singling out, somewhat arbitrarily, awardees who need neither a morale boost, nor the money – and to side with the Dodo in *Alice's Adventures in Wonderland* (and the 2020 Turner Prize judges) that 'everyone has won and all shall have prizes.'

My scepticism about prizes has been deepened by personal involvement – albeit in minor ways – in committees to select awardees. Even if there's a consensus about who did what, how much effort (or luck) was involved, and so forth, there can be divergences in the weighting of these criteria. Moreover, formal nominations are usually required before a candidate can be considered, so one may have regrets that one's most-admired candidates haven't been nominated.

I was taken further from my science-focused 'comfort zone' when I agreed to chair the judges for the Samuel Johnson Book Prize in 2013 – the UK's main award open to any category of non-fiction. The preliminary sifting, cutting down near-200 entries to about fifty, was rough and ready. There was a great deal of arbitrariness even at later stages (we could, for instance, have made up the entire shortlist from books on the First World War, of which there was a glut in the lead-up to the centenary year of its commencement). Our final choice of a single winner, a historical biography, was achieved by consensus – without any of the disputes or walk-outs that accompany the Booker Prize. But none of us, I suspect, would have been unhappy or surprised if our choice had coalesced about another highly rated shortlisted book on a very different theme – insects!

All the prizes I've mentioned aim to celebrate and reward

past achievements. But there is growing interest in 'challenge prizes'. These do the opposite: they incentivize future efforts to crack an important problem. The most prominent present-day exemplars are the 'X-Prizes', inspired by the Greek/American entrepreneur Peter Diamandis and run by his California-based foundation. Challenges are selected, and a jackpot of around $10 million is offered to those who first meet each of them. A special advantage of this system is that the aggregate funding expended by all the challengers for each prize far exceeds the prize money; each competition therefore offers a cost-effective incentive towards a goal that is socially worthwhile or of genuine public interest.

Challenge prizes have a long history. A prize to preserve food for Napoleon's army led to the invention of tinned food. A century later, another prize stimulated Lindberg's solo transatlantic flight. More recently there have been prizes for sub-orbital space flight, driverless cars, robots that operate in hazardous environments, and so forth. (And some haven't been won. Over a century ago a French foundation offered 100,000 francs for the first detection of extra-terrestrial life: moreover, finding it on Mars wouldn't count as it was deemed too easy!) As compared to usual forms of funding, these prizes encourage 'mavericks'. They can also attract public interest: those in robotics, for instance, can be a spectator sport where progress can be monitored and objectively measured.

But the most celebrated challenge prize dates from the early eighteenth century. In 1714, the British parliament passed the Longitude Act – establishing a fund of £20,000 (equivalent to several millions in today's money) for an effective method of measuring longitude at sea. This challenge led eventually to the rewarding of the Yorkshire carpenter and clockmaker John Harrison for his chronometer – one of the triumphs of eighteenth-century technology – which kept time to within 20 seconds after a transatlantic voyage on a rolling ship. Longitude is determined from the difference between local midday (inferred from when the Sun is highest in the sky) and the time of midday in Greenwich as recorded by the chronometer. (The Board of Longitude, incidentally, continued in existence for a century, funding instrumentation, polar exploration, etc.: it was in effect the first research council.)

Here I inject a personal perspective. The eight ex-officio members of the original Board of Longitude included the Astronomer Royal, the President of the Royal Society, and the Cambridge professor of astronomy. I've held these three posts (though that of Astronomer Royal, then the director of the Royal Observatory at Greenwich, is now an honorary one). In 2014 it seemed appropriate to commemorate the tercentenary of the Longitude Act, and I took the initiative in suggesting an anniversary Longitude Prize to address a current challenge. The idea was taken up by Geoff Mulgan, head of Nesta, a quango which already ran some challenge prizes. The government supported the initiative to the tune of £10 million. I chaired the advisory committee that came up with a shortlist of six topics where such prizes could make a difference.

We decided that the choice of challenge should involve the public and the media. There was a series of six BBC programmes each of which promoted a candidate project – followed by a public vote. The winning choice of challenge was to design an instrument that could, cheaply and quickly, identify whether a sick person's disease was bacterial or viral, and thereby avoid overuse of antibiotics against infections where they were ineffective. Antibiotic resistance is a mega-issue for those concerned about future global health. It had been raised on the agenda especially by Sally Davis, the UK government's former Chief Medical Advisor: she emphasized that we would enter a 'new dark age' of surgery if bacteria evolved immunity to standard antibiotics through their overuse. The challenge attracted more than 200 entries from around the world, whittled down later to about 30. It is being settled in September 2022, and will hopefully accelerate some innovations that could be hugely beneficial for the developing world. Many conventional research projects receive private or commercial funding; we hope that such sources might sponsor some of the other five challenges we shortlisted (involving equipment or software to help the paralysed or demented, new foods, or zero-carbon flight).

4.6 Sharing science

The anniversary Longitude Prize[10] managed to reach and engage a wide public much more quickly than was possible for the 1714 Longitude Act. Moreover, it attracted entries from around the world. This was of course due to modern-day communication technology.

Advanced technologies – computers and the web especially – offer huge benefits to today's young generation. But I think people as ancient as me had one compensating advantage. When our now-old generation were young, we could take apart a clock, a radio set, or motorbike, figure out how it worked, and then reassemble it. And that's how many of us got hooked on science or engineering. Going back further, the young Isaac Newton made model windmills and clocks – the high-tech artifacts of his time. Darwin collected fossils and beetles. Einstein was fascinated by electric motors in his father's factory.

It's different today. The gadgets that now pervade our lives, smartphones and suchlike, are baffling 'black boxes' – pure magic to most people. Even if you take them apart, you'll find few clues to their arcane miniaturized mechanisms. And you certainly can't put them together again. So the extreme sophistication of modern technology, wonderful though its benefits are, is ironically an impediment to engaging young people with basics – with learning how things work. Likewise, town-dwellers are more distanced from the natural world than earlier generations were. And many children never see a dark sky – or a bird's nest.

But one shouldn't be nostalgic. These losses are, arguably, far outweighed by the upside. Even if children can't see wild creatures 'for real', there's huge fascination in wildlife films portraying the variety and wonder of the natural world; experiments, and natural events such as tropical storms or even the impact of a comet on Jupiter, can be followed in real time by anyone who is interested. AI can't adequately replace a real teacher; but can offer valuable supplementation, as well as personalized help that a single teacher couldn't adequately supply to a large class.

Moreover, the internet allows new styles of research in which amateurs, young and old, can participate. A pioneering example was the Galaxy Zoo project, instigated by Chris Lintott at Oxford.[11] Images of 3 million galaxies could be viewed on the web, and the labour-intensive task of classifying them was shared by thousands of keen amateur astronomers. And among other projects, there was a programme to digitize weather reports from eighteenth-century ships' log-books – labour-intensive, but interesting for climate science. This had the bonus of stimulating broader interest in naval history among some participants. The Washington-based Center for Open Data Enterprise (CODE), established in 2015, aims to encourage governments, businesses and researchers to make data available and to facilitate its use by others who can distil information from it.

More surprisingly, wiki-style activity even has scope in mathematics. On the 'Weblog' of Cambridge professor Tim Gowers,[12] theorems have been proved via a genuine collective effort, like completing a jigsaw, or the development of open-source software. And if you seek expert commentary, you can – to take just two near-random examples – turn to blogs and articles by Scott Aaronson[13] for quantum computing; and to Ed Yong on biological topics. Many blogs are an authoritative match for what you'll find in hard covers. These are just instances of how scientific progress and enlightenment is being enhanced by the involvement of millions of people worldwide – and by easier access to the views of leading experts (albeit embedded in the junk that dominates the internet).

Some pessimists argue that scientific progress will clog up because of 'information overload'. I don't think that's a serious worry. Novel advances bring with them a flood of new data, but the cost of storing and processing vast datasets is falling towards zero. For instance, the European space observatory GAIA gathered data on nearly 2 billion stars. Computerized analysis of this huge sample reveals patterns and regularities, which cut down the number of disconnected facts worth remembering. There's no need to record the fall of every apple, because, thanks to Newton, we understand how gravity pulls everything – whether apples or spacecraft – towards the Earth. But there is of course a need to distil the essence of new work so that it can be absorbed by individual researchers. Here the role of formal review articles, and more informal blogs, is of growing importance.

People everywhere in the world are immersed in a cyber-space that is ever more information-rich and sophisticated. Moreover, students or scholars can access large datasets – from a 'virtual observatory' or from a library of genome data. They no longer need to go to a central archive, any more than scholars actually need to visit a great library (except to study actual manuscripts). The Royal Society's journal founded by Henry Oldenberg was the prototype for the tens of thousands of refereed journals that exist today. Printed journals were a real advance in the 1660s. But they are now anachronistic – so indeed are expensive hard-cover mono-graphs. Putting journals online is cheaper and offers greater ease in tracking references and data through keywords, links, and so on. But ideally scientific information and ideas should be freely available. We're closer to the ideal in some subjects than in others: understandably, tensions are stronger when the work is patentable and commercially exploitable.

Researchers in physics and astronomy across the world post papers on a web archive[14] established by Paul Ginsparg, now at Cornell University – and read it daily. Most papers later appear in traditional journals; but that is for accredita-tion, not to get more readers. It's rather sad that thanks to Paul Ginsparg the educated public has been able for dec-ades to read everything on, for instance, superstring theory – which will not enlighten them much – but cannot freely access writings in the humanities.

Going further, it is unclear whether the scientific paper will remain the preeminent 'publishable unit' for much longer. The time for processing, reviewing, revising and eventually publishing an academic article stretches over many months – indeed years in some fields. It is surely constricting that the career prospects of young academics depend on a single monograph, or on the bibliometric scores of a few papers. It is even worse if there is an institutionalized pecking order of journals and they have to struggle for acceptance in a top-ranked journal. One of the most deplorable remarks I heard recently was a professor responding to my question 'how do you decide on whether a paper is good?' with the reply 'By the journal it's in.' Moreover, it is sad if university administrators, motivated to boost their institution's rating in league tables, pressurize staff to publish in American journals simply to garner higher citations. (And we should remember how all these indicators unfairly favour those whose native language is English.)

Even the refereeing role of journals may one day be trumped by more informal systems of quality rating. Already, some web archives allow readers to write comments or brief reviews. Provided these were not anonymous, even an indication of 'likes' or 'stars' from highly regarded experts would be a plus for the author and a guide to readers bewildered by the volume of material on offer. Blogs and wikis will play a bigger role in critiquing and codifying science. The legacies of Gutenberg and Oldenberg are not optimal in the age of Zuckerberg.

Today we've moved away from the 'two cultures' of the arts and sciences. There's at least a 'third culture', embracing the social sciences; indeed, today it would be truer to say that 'culture' has many interweaving strands. Nonetheless, intellectual narrowness and ignorance are still endemic; and there are worryingly many, especially in influential positions in politics and the media, to whom the sciences remain a closed book. This remains, incidentally, a special concern in England, where young people are forced towards a specialized curriculum by the age of 16. To broaden everyone's perspective – and of course to keep career options open – is an imperative. It's a campaign where I've made common cause with my 'humanistic' counterparts at the British Academy.

The humanities and social science should engage us as human beings. But scientists have an extra reason for supporting their continuing prominence in our universities – for promoting not just science, technology, engineering and mathematics, but the arts (broadly interpreted) as well: STEAM, not just STEM. These subjects sensitize and guide the public, as citizens, in assigning how science should be applied. There's an ever-widening gap between what science allows us to do and what it is prudent or ethical actually to do. This book has emphasized widely held anxieties that genetics and artificial intelligence may 'run away' too fast – and that our imprint on the global environment could be irreversibly damaging. Answers to such dilemmas cannot come from within natural science itself, but as part of their education students should be attuned to these issues.

I conclude, therefore, with some further thoughts on tertiary (post-18) education and its links to research – while reiterating that improvements in pre-18 and technical education deserve at least equal priority.

4.7 Enhancing science education

Most university campuses around the world were silent and deserted during the peak of Covid-19. Their life was gradually restored. But nobody expects full reversion to the 'old normal' – nor should we wish for it. The recent crisis should energize and accelerate much-needed reforms to the whole post-18 education sector. It needs to become more flexible and open, recognizing that what young people learn today will need regular updating through their lives – as fast-changing lifestyles offer new opportunities for both work and leisure.

There's now more experience of online and remote teaching: we can make a more realistic assessment of what's the most effective use of 'contact' hours with students. We can also learn from institutions that had already spearheaded that transition pre-pandemic – for instance the fast-expanding Arizona State University. In more traditional universities, the basic lectures on core topics are given 'live' to audiences of 200+. There's no real feedback or discussion during these performances – though they are, at least in the better institutions, supplemented by smaller classes and tutorial groups, often run by 'teaching assistants'. Little would be lost if typical big lectures were videoed rather than 'live'. Indeed, they could then be more carefully prepared and achieve higher quality. Moreover, not only could they be watched more than once by their primary student audience, but they could be disseminated globally. There have been successful precedents at MIT and Stanford. Scholars such as Harvard's Michael Sandel have become international 'stars'. The wider viewing of excellent lectures (even if not part of a course offering online credits) should surely be welcomed as a contribution to public understanding.

What's needed, however, is more than just incorporation of virtual activities into the existing framework. Students should be able to choose their preferred balance between online and residential courses (and, as mentioned, to access distance learning of higher quality). Purely online courses, the so-called MOOCs (massive open online courses), have had an ambivalent reception. As stand-alone courses without complementary contacts with a real tutor, they are probably only satisfactory for masters-level vocational courses intended for motivated mature learners studying part time. But they can have wider benefits as part of a 'package' that incorporates 'live' tutoring as well. We need more facilities for part-time study and lifelong learning, and a blurring of the damaging divide between technical and university education.

Students who realize that the course they embarked on isn't right for them, or who have personal hardship, should be enabled to leave early with dignity, with a certificate to mark what they've accomplished. They shouldn't be disparaged as wastage: they should make the positive claim that 'I had two years of college.'

And, more important, in adapting to a world of greater mobility, the higher and further education system should enable everyone to have the opportunity to re-enter higher education (maybe part-time or online) at any stage in their lives. This path could become smoother – indeed routine – if there were a formalized system of transferable credits across the whole system of further and higher education. There should be a flexible grant or loan system offering entitlement to three years' support, to be taken 'à la carte', year by year or via a succession of 'modules', at any stage in life. This would mean, for instance, that those who didn't complete an undergraduate course when young could get some certificate of credit and an entitlement to return and 'upgrade' later. The aim should be to leave options open so that people's lives aren't constrained by what they have achieved (or failed to achieve) by the age of 21. These opportunities promote social mobility for individuals. But an important corollary is the benefit of these reforms to society as a whole. These reforms would clearly offer greater opportunities and more varied career paths to the students. But we all benefit from interaction and collaboration with people from different walks of life, with different technical experience, who can blend the two cultures, who can straddle science, public policy and the media, or who have experience working with communities that would benefit specially from the application of particular scientific knowledge – or, conversely, might be specially at risk from it.

A feasible modest step that would send an encouraging signal would be for universities whose entry bar is dauntingly high to reserve a fraction of their places for students who don't come straight from school – thereby offering a second chance to those who were disadvantaged at 18, but have caught up by earning two years' worth of credits at other institutions or online. Such students could then advance to degree level in two further years.[15] In the US, the University of California system was conceived by Clark Kerr so that it provided this flexibility. Only about half the students joining the elite campus at Berkeley transfer directly from High School; others come via Junior or Four-Year colleges.

But it's a sad fact that the worst educational inequalities are imprinted earlier in life – in the infant and pre-school years, and during school education. It will be a long slog to ensure that high-quality teaching at school is available across the full geographical and social spectrum – indeed, this may be impossible until there is a narrowing of the gulf between the resources of the private fee-taking schools and those in the state system. However, promoting life-long and part-time learning, with flexible assessment, would go some way to offering more support to those whose deprivations start in infancy, and lead to barriers that become harder to surmount, and to exclusions that offer no second chances.

Restructuring is specially needed in the UK's universities, where there's currently a systemic weakness: their missions are not sufficiently varied. They all aspire to rise in the same league table, which gives undue weight to research rather than teaching. Most of their students are between 18 and 21 – undergoing three years (or maybe four) of full-time (generally residential) education and studying a curriculum that's too narrow, even for the minority who aspire to professional or academic careers. Even worse, the pre-university school curriculum is too narrow as well. The campaign for an international baccalaureate-style curriculum for 16–18-year-olds has been impeded by universities, whose entrance requirements overtly disfavour applicants who straddle science and humanities – precisely the sort of crossover we need as a society to effectively implement science to tackle the challenges introduced in chapter 1, and others around the corner of which we may as yet have no knowledge.

It is plainly no longer possible to be a polymath, as it was before the twentieth century; a broad curriculum is, however, surely desirable at school and in the early university years. But, of course, graduate-level education, crucial for professional scientists, is inherently specialized. In the US, only a minority of universities have strong graduate schools. That is a model the UK should move towards. Many who teach in the best American liberal arts colleges (such as Haverford, Wellesley, and William and Mary), which offer broad and top-rated undergraduate education but not PhD courses, are productive researchers and scholars – it's important to emphasize that concentration of graduate education need not, especially in the humanities, entail an equivalent concentration of research (a distinction that is often conflated). But if these scholars have graduate students, the students are based in another university. The key point is that, in some fields at least, a researcher can do distinguished work alone. But students aspiring to a PhD need more than just a good supervisor. They need to be in a graduate school where courses are offered over a wider range. Without this second component, a newly minted PhD will not necessarily have the flexibility and range that is needed for their later career – whether in the public or private sector.

There are in any case strengthening reasons to query the value of the PhD in its present form. It's regarded as an 'entry ticket' to academia – but only a minority of those taking the degree have this aim: many end up in industry, or in the public sector – often in roles crucial to solving the global challenges explored in chapter 1, which require a wider range of expertise than just writing papers for scholarly journals. A wider-ranging (or longer but part-time) postgraduate programme might be a more stimulating and relevant experience for people who are not seeking an academic career but would undoubtedly benefit from contact with scholars and scientists. Indeed, such a programme would likely be more stimulating and relevant for would-be academics, too. The system should be flexible enough to cope with a world where most of us will shift careers and seek updated expertise throughout our working lives.

Inevitably, relatively few universities attract the lion's share of research funding – and have graduate schools spanning all faculties. That is likely to be true whatever system prevails. But despite the trend towards concentration, I think it is crucial to avoid formalizing the hierarchy, and to retain a system that allows research excellence to sprout and bloom anywhere in the system. For instance, Leicester University in the UK is a leading centre in space science and genetics. This is largely because the funding system was flexible enough to fund the ambitions of two young faculty members in the 1970s – the pioneer space-astronomer Ken Pounds, and Alex Jeffrey, inventor of DNA fingerprinting. We should surely be encouraged by this example where researchers were trusted and (calculated) risks were taken with funding.

Let's hope that the current crisis catalyses constructive innovations in higher education worldwide. This sector – crucial to our future – mustn't be sclerotic and unresponsive to changes in needs, lifestyle and opportunities. A rethink is overdue – indeed nations cannot afford to be complacent about the present structures. The system should enable everyone to refresh or upgrade their expertise throughout their lives; and should offer online material freely to all who wish to engage simply through interest, rather than in a quest for qualifications.

4.8 On ivory towers

I conclude this chapter with some thoughts that may seem parochial and self-indulgent: about Oxford and Cambridge – which loom unhealthily large in the UK but are the institutions I know best.[16] They are routinely compared to the Ivy League, but that is a bit misleading. Their balance of public and private funding is closer to the top state universities in the US – Berkeley, Michigan or Texas. But what renders them unique in the world is that – thanks to the college system – they combine the strength of world-class research universities with the pastoral and educational benefits of the best American liberal arts colleges. That's why, according to a report from the Higher Education Policy Institute (HEPI), their students show a higher satisfaction rating, and work harder, than those studying elsewhere in the UK.[17] Incidentally, HEPI found little correlation among the rest of the UK universities between student satisfaction and placing in 'league tables'. This isn't surprising, because these tables focus on research. Indeed, Oxford and Cambridge would rank even higher in international comparisons if the 'league tables' gave proper weightings to teaching and the student experience, rather than having a focus on research, especially in science.

Why have these ancient institutions succeeded, despite fast-changing times? To a typical business consultant, their organograms look nightmarish – an intricate matrix of colleges and departments. Although this has its downsides, it offers genuine advantages over a cleaner system of line management. Academics feel less pushed around. As their careers develop, they can find a niche, an optimum individual mix of teaching, research and administration.

Universities need to be business-like; so does a hospital, so even does a church. But that does not mean they should be like a business – indeed the inchoate 'partnership' model is remarkably cost-effective. It is through this flexibility that Oxbridge retains the dedicated institutional loyalty of hundreds of highly able (and opinionated) people despite very un-stratospheric financial rewards. And the flexible structure has undoubtedly helped to foster the innovative culture that has created, in both Oxford and Cambridge, the creative hubs of labs and start-ups described earlier.

And, of course, neither is in any sense a technical university: the traditional arts and humanities continue to flourish and to leaven the scientific scene. In that context I offer a fogeyish though not entirely frivolous reminiscence that's perhaps relevant in an era where the humanities are under-appreciated. Some years ago I spoke at a dinner hosted by Oxford's Vice-Chancellor to honour entrepreneurial academics who'd gained big and impactful grants. I reminded them that two of the most valuable pieces of intellectual property to come from Oxbridge did not come from scientists or engineers – but from Cambridge's Professor of Renaissance Literature and Oxford's Professor of Anglo-Saxon. I refer of course to C. S. Lewis and J. R. R. Tolkien; their works now, decades later, earn billions for the so-called creative industries. These two eminent scholars – both, in style and attitudes, archetypal old-style Oxbridge dons – would feel disaffected aliens in today's world of line management, and the audit culture. Their values were the traditional ones: commitment to an institution, and to scholarship and learning for their own sake. We'll all be losers if there weren't at least a few universities that continued to nourish such people.

Oxbridge shouldn't dwell unduly on past glories, but there's no gainsaying the extraordinary global impact that has stemmed from discoveries and insights – spanning all the sciences and humanities – that germinated on the small patches of ground where these universities stand, and where so many influential figures spent formative years. The global expansion of education in a high-tech world will surely create new models, taking advantage of greater mobility and new technology. But ancient universities will still have their place, as beacons of continuity in a runaway world.

Afterword

Over the last two decades, everyday life, and the world's economy, have been transformed by new technologies: the internet, robotics, gene-editing, clean energy and so forth. They will prove even more transformative in the coming years.

I began this book by highlighting in chapter 1 some hopes and fears: these technologies – and new ones opened up by future scientific advances – could, if optimally deployed, have utopian consequence; but in contrast, they could lead to a new dark age. The stakes are getting higher. Indeed, this is the first era in which humanity is sufficiently empowered and dominant to affect our planet's entire habitat: the climate, the biosphere and the supply of natural resources.

Our world is increasingly interconnected, reliant on globe-spanning networks for food supply and manufacturing. Novel viruses can emerge unpredictably at any time and spread with devastating speed. And we're now more mindful that our fragile and interconnected society is vulnerable to other scenarios – massive cyber attacks, cascading failures of crucial infrastructure or accidental nuclear war – whose likelihood and impact are rising year by year. We're developing new technologies that offer huge benefits but also create new ways for humanity to harm itself – via the consequences of error or terror.

As science's potential becomes more powerful and pervasive, it's ever more crucial to ensure that it is deployed optimally, and that the brakes are applied to applications that are dangerous or unethical. But we must expect technology-driven changes in lifestyles – in the nature of work, education and social interactions – even though we can't predict the innovations that will drive them. These will happen on a timescale of decades, playing out against a shifting backdrop of political tensions between the 'West', China and the Global South. This is more rapid than the overall societal changes that occurred in earlier centuries; on the other hand, it is slow enough to give nations time to plan a response: to mitigate or adapt to a changing climate, to modify lifestyles and to achieve sustainable modes for food and energy production. Such transformations are possible in principle – most of the relevant science is already known – though there is a depressing gap between what is ethically and humanly desirable and what actually occurs. Let's hope that the Covid-19 crisis will change mindsets towards a greater awareness that 'we are all in this together', with more focus on a 'global' perspective. There are a few rare times when there seems special motivation to focus on the prospects for all of humanity. This feels like one of those times.

Science has utterly transformed our world. Later chapters in the book have focused on the role of scientists in the broader societal and political context. The prime goal of science is to understand our physical and biological environment in all its complexities. It has enhanced our perception of nature, and thereby removed some of the irrational dread suffered by our medieval forebears. But its applications have created a world where baffling and sometimes scary artefacts pervade everyday life.

In the days when newspapers and public radio were most people's prime sources of news, extreme and 'fake' claims were muffled by responsible journalists. But in the domain of the internet, such claims are amplified by 'clicks' to still more extreme views. That's why I've emphasized the special obligations of scientists themselves. They should do all they can – through involvement in education and engagement with the media – to ensure that the wider public has a balanced view. And they have a responsibility to ensure that, when their findings can lead to practical innovations, these are beneficial; they should speak out (and, when appropriate, inform their governments) against potentially unethical or dangerous applications. But, when they stray beyond their specific field of expertise (as individuals or via other professional bodies), it's important to recognize that they speak 'only' as 'concerned citizens'.

I've emphasized that everyone's education should help them understand, at least in outline, the world that present-day science has revealed and helped to shape. Otherwise they can't appreciate nature's wonders and mysteries, nor feel at ease with the high-tech systems that pervade their lives. And science is a truly global culture. But this education is also a prerequisite for genuine participation, as informed citizens, in discussions and democratic decisions on issues with a scientific dimension – for instance energy, health and the environment. Everyone needs enough 'feel' for science to realize that (as was clear in the pandemic) there are crucial uncertainties, but that nonetheless the views of scientists who are striving to reduce these uncertainties deserve special weight. And, of course, they need enough numeracy to avoid being flummoxed by statistics and enough scepticism to ignore manifestly flaky internet sites.

So how can we ensure a safer world? No single country has the power to prevent risks such as pandemics or climate change on their own. Global issues require global cooperation. It's deeply imprudent for nations to ignore potentially catastrophic scenarios, and not to prioritize precautions to minimize the risks they pose. Threats such as pandemics could cost trillions of dollars as well as millions of lives. So it is worth spending far more to ensure preparedness and resilience that can reduce the likelihood and impact of such nightmarish events. We depend increasingly on global infrastructures – the internet, GPS and so forth – and it is crucial to ensure that these are robust. Likewise, events in 2022 have highlighted the vulnerability of the supply chains for energy and food – and how interdependent all nations are.

Nations may need to cede sovereignty to more international bodies like the WHO to regulate dangerous technologies and minimize catastrophic risks. There seems no scientific impediment to achieving a sustainable world beyond 2050, in which the developing countries have narrowed the gap with the developed, and all benefit from further advances as important as that of vaccines and information technology in the last decade. But the politics and sociology engender pessimism. Will richer countries recognize that it's in their self-interest for the developing world to prosper, sharing fully in the benefits that science offers? And can the focus of our sympathies become more broadly international? Can nations sustain effective but non-repressive governance in the face of threats from small groups with high-tech expertise? And, above all, can our institutions prioritize projects that are long term in political perspective even if a mere instant in the history of our planet?

The answers will depend on public attitudes. International coordination will be crucial in narrowing the gap between the way the world is, and the way it could be if science were optimally applied. Democratic politicians will support this goal, and take a long view, if the public supports them. For this to happen, we need a scientifically aware public energized and inspired by charismatic campaigners. We have seen, for instance, what Pope Francis, David Attenborough, Bill Gates and Greta Thunberg have achieved in shifting public perspectives on crucial long-term goals. We need more such individuals to influence us and our political leaders – individuals who resonate with science, but can inspire the ethical guidance and motivation that science alone can't offer.

To quote an optimistic thought from Margaret Mead,

> Never doubt that a small group of thoughtful, committed citizens can change the world; indeed, it's the only thing that ever has.

Notes

Chapter 1: Global Mega-challenges

1 Ehrlich, P. R. 1968. *The Population Bomb*. New York: Ballantine Books.

2 Meadows, D. H., D. L. Meadows, J. Randers and W. W. Behrens. 1972. *The Limits to Growth*. New York: Universe Books.

3 The UN 'World Population Prospects' quotes a best estimate of 9.7 billion for the world's population by 2050. Another authoritative source is the Population Project of the International Institute for Applied System Analysis (IIASA) which quotes somewhat lower figures. Recent data on the fertility rate in India may lower these projections further.

4 See https://data.worldbank.org/topic/poverty

5 See, for instance, Pretty, P. and Z. P. Bharucha. 2014. 'Sustainable intensification in agricultural systems', *Annals of Botany*, 114: 1571–96.

6 Every year the WWF reports 'world overshoot day' – the day in the year after which our demand on nature exceeds Earth's annual biocapacity. In 2022 this day was 28 July.

7 The initial reference is Rockström, J. et al. 2009. 'Planetary

boundaries: exploring the safe operating space for humanity', *Ecology and Society*, 14: 32. It is updated by the Stockholm Resilience Centre at https://www.stockholmresilience.org/re search/planetary-boundaries.html

8 This quote appears in the preface to the Dasgupta review (see note 9).

9 Dasgupta, P. 2021. *The Economics of Biodiversity: The Dasgupta Review*. London: HM Treasury, https://www.gov .uk/government/publications/final-report-the-economics-of -biodiversity-the-dasgupta-review

10 Stern, N. 2006. *The Economics of Climate Change: The Stern Review*. London: HM Treasury, https://www.lse.ac.uk/grant haminstitute/publication/the-economics-of-climate-change -the-stern-review/

11 Wilson, E. O. 2006. *The Creation: An Appeal to Save Life on Earth*. New York: W. W. Norton.

12 World Commission on Environment and Development (Brundtland Commission). 1987. *Our Common Future: Report of the World Commission on Environment and Development*. Oxford: Oxford University Press.

13 For history of the curve and updates, see https://keelingcurve .ucsd.edu

14 IPCC. 2021. *Climate Change 2021: The Physical Science Basis. Contribution of Working Group I to the Sixth Assessment Report of the Intergovernmental Panel on Climate Change*. Cambridge: Cambridge University Press, https://www.ipcc .ch/assessment-report/ar6

15 Robinson, E. 2021. 'Opinion: The UN's dire climate report confirms: We're out of time', *The Washington Post*, 9 August, https://www.washingtonpost.com/opinions/2021/08/09/uni ted-nations-climate-report-dire

16 A further concern, given the complexity of the interactions, is that if the changes exceed some threshold – a 'tipping point' – there could be no expectation that the climate would revert

to its pre-industrial state even if the excess CO_2 were to be eventually sequestrated.

17　Bishop, B. 2021. 'National Ignition Facility experiment puts researchers at threshold of fusion ignition', Lawrence Livermore National Laboratory website, 18 August, https://www.llnl.gov/news/national-ignition-facility-experiment-puts-researchers-threshold-fusion-ignition

18　The concept was re-launched at COP26 in Glasgow: see http://mission-innovation.net/about-mi/overview

19　Although first published in 2016, Oliver Morton's book *The Planet Remade: How Geoengineering Could Change the World* (Princeton: Princeton University Press) remains an excellent summary.

20　This was published in 1924 as a hardback but is available online at https://www.marxists.org/archive/haldane/works/1920s/daedalus.htm

21　Two highly accessible books on these developments are: Doudna, J. A. and S. S. Sternberg. 2017. *A Crack in Creation*. Boston, MA: Houghton Mifflin Harcourt (Jennifer Doudna is one of the inventors of CRISPR/Cas9); and Mukherjee, S. 2016. *The Gene: An Intimate History*. New York: Scribner. See also Isaacson, W. 2021. *The Codebreaker*. New York: Simon and Schuster.

22　See also the description of work in 1918 on the horsepox virus in Noyce, R. S., S. Lederman and D. H. Evans. 2018. 'Construction of an infectious horsepox virus vaccine from chemically synthesized DNA fragments', *PLOS One*, 13(1): e0188453.

23　The website is https://longbets.org

24　Steven Pinker and I wrote a joint article on this in the 16 June 2021 issue of the *New Statesman*.

25　The reference is https://thebulletin.org/2021/05/the-origin-of-covid-did-people-or-nature-open-pandoras-box-at-wuhan. Further work has led to controversy, and the suspicion that the origin of Covid-19 will remain mysterious.

26 Carter, S. L. 2021. 'If Covid did escape from a Wuhan lab, brace yourself: the world's anger will be terrible to behold', Bloomberg website, 4 June, https://www.bloomberg.com /opinion/articles/2021-06-04/if-covid-did-escape-from-a-wuhan-lab-brace-yourself

27 For example: Hoffman, B. J., M. K. Shoss and L. A. Wegman (eds). 2020. *The Cambridge Handbook on the Changing Nature of Work*. Cambridge: Cambridge University Press; Susskind, D. 2020. *A World Without Work*. London: Penguin; Lee, K.-F. 2018. *AI Superpowers: China, Silicon Valley and the New World Order*. New York: Houghton Mifflin; Frey, C. B. 2020. *The Technology Trap: Capital, Labour and Power in the Age of Automation*. Princeton: Princeton University Press.

28 This particular claim, made in 2011 on the basis of data from Israeli courtrooms, has been queried because the trend could have been due to the custom of scheduling, before the 'break', cases likely to be quickly settled because the defendant had no legal representation. But there is no gainsaying the multiple evidence for judicial biases.

29 Bostrom, N. 2014. *Superintelligence: Paths, Dangers, Strategies*. Oxford: Oxford University Press.

30 Tegmark, M. 2017. *Life 3.0*. New York: Knopf/Allen Lane.

31 Derek Parfit's arguments are presented in part 4 of his *Reasons and Persons* (New York: Oxford University Press, 1984).

32 Diamond, J. 2005. *Collapse: How Societies Choose to Fail or Succeed*. New York: Penguin.

Chapter 2: Meet the Scientists

1 This term was introduced and defined by T. S. Kuhn in *The Structure of Scientific Revolutions* (Chicago: University of Chicago Press, 1962). Another classic book influential among scientists has been Karl Popper's *Logic of Scientific Discovery* (London: Routledge, 1959) – a translation of the original

German version published in 1934. In the intervening years, Popper enhanced his reputation with his deeply impressive contribution to political theory: *The Open Society and Its Enemies* (London: Routledge, 1945). The accessible book *The Meaning of Science*, by Tim Lewens (New York: Basic Books, 2016), offers a clear critique of the viewpoints of Popper, Kuhn, and others.

2 The tone of much criticism is exemplified by this article: Schneider, L. 2017. 'Human Brain Project: bureaucratic success despite scientific failure', For Better Science website, 22 February, https://forbetterscience.com/2017/02/22/human -brain-project-bureaucratic-success-despite-scientific-failure

3 Darwin, C. 1860. *On the Origin of Species by Means of Natural Selection, or the Preservation of Favoured Races in the Struggle for Life*. London: John Murray.

4 *A Short History of Nearly Everything* (London: Doubleday, 2003) and *The Body: a Guide for Occupants* (London: Doubleday, 2019).

5 University of Michigan. 2021. 'Evolution now accepted by majority of Americans', ScienceDaily website, 20 August, https://www.sciencedaily.com/releases/2021/08/210820111042.htm

6 Howes, A. (2021) 'Upstream innovation: Raising the status of innovation and innovators', in *The Way of the Future*, published by the Entrepreneurs Network and the Blair Institute, a think-tank founded by Tony Blair, former Prime Minister of UK.

7 The ITER apparatus, intended to explore the feasibility of commercial-scale nuclear fusion (see section 1.2) has a projected cost more than twice that of the LHC or the JWST.

8 An excellent account of this discovery and its context is given in Schilling, G. 2017. *Ripples in Spacetime*. Cambridge: Belknap Press of Harvard University Press.

9 Weinberg, S. 1994. *Dreams of a Final Theory*. New York: Vintage Books.

Chapter 3: Science Comes out of the Lab

1 It was not transmitted by a virus but was a 'prion disease'. Such diseases, also known as transmissible spongiform encephalopathies or TSEs, are a group of rare, fatal brain diseases that affect animals and humans. They are caused by an infectious agent known as a prion, which is derived from a misfolded version of a normal host protein known as prion protein.

2 See Nature Editorial. 2022. 'Eric Lander's resignation for bullying raises questions for the White House', *Nature*, 602: 361–2.

3 An excellent biography is Brown, A. 2012. *Keeper of the Nuclear Conscience: The Life and Work of Joseph Rotblat.* Oxford: Oxford University Press.

4 Another US vehicle for engaging independent scientists with defence issues – this one focused primarily on nuclear arms control and non-proliferation – is the US National Academy of Sciences permanent Committee on International Security and Arms Control (CISAC). It was formed in 1980 to draw on the top talent from the US National Academies. It has long engaged in bilateral discussions with similarly constituted counterpart groups in Russia (since 1981), China (since 1988) and India (since 1999).

5 Garwin's remarkable and sustained career (he was born in 1928) is described in Shurkin, J. N. 2017. *True Genius: The Life and Work of the Most Influential Scientist You've Never Heard Of.* Buffalo, NY: Prometheus.

6 CSER's website is https://www.cser.ac.uk

7 The Oxford Martin School's website is https://www.oxfordm artin.ox.ac.uk

8 The text of the encyclical: Pope Francis. 2015. *Laudato Si': On Care for Our Common Home.* Vatican City: Libreria Editrice Vaticana, can be found at https://www.vatican.va/content/ francesco/en/encyclicals/documents/papa-francesco_ 20150524_enciclica-laudato-si.html

9   Juncker's remark is quoted in *The Economist*, 15 March 2007.
10 See this website for a description of these issues: https://lordslibrary.parliament.uk/government-investment-programmes-the-green-book
11 The work of the IMF on these figures can be found at https://www.imf.org/en/Publications/SPROLLs/covid19-special-notes
12 In contrast, there is still no effective anti-HIV vaccination, after 40 years.
13 Information on the Royal Society can be found at https://royalsociety.org
14 All went smoothly at the anniversary event, except that a faulty sound system, though fine for the audience, rendered the speeches distorted and inaudible to those on the stage. The 'royals' therefore had even more excuse for looking bored than they would normally have done on such occasions.
15 See website at http://www.nasonline.org
16 See https://council.science/
17 The Pontifical Academy of Science has a website giving full information about its membership and reports of its plenary sessions, conferences and statements: https://www.pas.va
18 This is the resultant report: Select Committee on Risk Assessment and Risk Planning. 2021. *Preparing for Extreme Risks: Building a Resilient Society.* London: House of Lords, https://committees.parliament.uk/publications/8082/documents/83124/default/

Chapter 4: Getting the Best from Science

1   Wikipedia offers a good summary of graphene's discovery, properties and potential applications: https://en.wikipedia.org/wiki/Graphene
2   The future relationship with ESA is actually less straightforward than with ESO and CERN. The UK continues its mandatory contribution to ESA's space science, but ESA

receives direct funding from the European Union for the Copernicus Earth-observation programme and for the Galileo system (a planned European counterpart to GPS).

3 The OECD oversees the PISA tests of attainment levels of 15-year-olds in different countries: https://www.oecd.org/pisa

4 See https://twas.org

5 The data on NIH grant statistics for the last 20 years can be found here: Lauer, M. 2021. 'Long-Term Trends in the Age of Principal Investigators Supported for the First Time on NIH R01-Equivalent Awards', National Institutes of Health website, 18 November, https://nexus.od.nih.gov/all/2021/11 /18/long-term-trends-in-the-age-of-principal-investigators -supported-for-the-first-time-on-nih-r01-awards

6 This classic document is reprinted here: http://www.educ ationengland.org.uk/documents/robbins/robbins1963.html

7 This article summarizes the near-100-year history of this laboratory: https://en.wikipedia.org/wiki/Bell_Labs

8 See Kirk, K. and C. Cotton. 2016. *The Cambridge Phenomenon: 50 Years of Innovation and Enterprise*. London: Third Millennium.

9 https://www.statista.com/statistics/1096928/number-of-glo bal-unicorns-by-country

10 For details of these awards, see https://longitudeprize.org

11 https://www.zooniverse.org/projects/zookeeper/galaxy-zoo

12 Tim Gower's 'Weblog' is a great resource for mathematics: https://gowers.wordpress.com

13 https://scottaaronson.blog

14 The archive is at https://arxiv.org. New and separate websites have been set up to cover biological topics where the 'pre-print' tradition took longer to develop.

15 To persuade colleagues in my own ancient institution, Trinity College, Cambridge, to accept this reform – despite traditional reluctance to 'do anything for the first time' – I

would remind them that they would actually be reviving a successful tradition. In the nineteenth century two of our greatest alumni – the astronomer Arthur Eddington, and J. J. Thompson, discoverer of the electron – both transferred from Owens College, Manchester, and achieved top honours in their mathematics degree two years later.

16 I spent most of my career in Cambridge, as a researcher, professor and institute director, feeling fortunate to be based there, for the reasons summarized in this section. So, despite reservations, I didn't give a straight 'no' when asked if I might wish to be considered as Master of Trinity (Cambridge's largest college). Some earlier Masters had found the College difficult and contentious: I worried I might be joining the 'Unholy and Divided Trinity', but I was friendly with the retiring Master, Amartya Sen, and with his predecessor, Michael Atiyah. These men, each with a record of academic distinction to which I couldn't aspire, were both entirely unstuffy and progressive in their views – and they were encouraging. So I took on this non-executive role in 2004, staying until 2012. My 'watch' fortunately proved generally uncontentious, and indeed deepened my loyalty to the university.

17 Data are in this report: Chester, J. and B. Bekhradnia. 2009. *How different are Oxford and Cambridge?* Oxford: Higher Education Policy Institute, https://www.hepi.ac.uk/wp-con tent/uploads/2014/02/44-Oxford-and-Cambridge-summa ry.pdf. This report gives fuller data on satisfaction levels of various categories of students in all UK universities in 2020 (satisfaction understandably plummeted in 2021 because of the pandemic): Neves, J. and R. Hewitt. 2020. *The Student Academic Experience Survey 2020.* Oxford: Higher Education Policy Institute, https://www.hepi.ac.uk/wp-content/uploads /2020/06/The-Student-Academic-Experience-Survey-2020. pdf

Index